PLC Programming Using RSLogix 500

A Practical Guide to Ladder Logic and the RSLogix 500 Environment

Nathan Clark

Books in this Series

PLC Programming
Using RSLogix 500

A Practical Guide to Ladder Logic
and the RSLogix 500 Environment

PLC Programming
Using RSLogix 5000

Understanding Ladder Logic
and the Studio 5000 Platform

Table of Contents

1. Introduction to RSLogix 500 and PLCs

This PLC programming and technical guide will familiarize you with the RSLogix 500 interface, methods to build a PLC program, and how to connect to a MicroLogix PLC. This book will help you start from scratch and be able to design and build a program in RSLogix 500. The focus software platform of this guide is RSLogix 500, but we will briefly cover RSLinx, and FactoryTalk Studio. This guide does however not cover how to select hardware, or what PLC is best for the job. In this chapter we will touch on the RSLogix software, compatible PLCs, and go over what you can expect to learn from this guide in more detail.

1.1 Intended Audience

This technical guide is intended for beginners with no prior PLC programming experience. However, we do assume the following:

- You are familiar with basic control equipment (such as push buttons, position sensors, switches etc.)

- You already have RSLogix 500 downloaded

- You already have FactoryTalk Activation Manager installed, and all proper licensing is in place

1.2 Important Vocabulary

- PLC = Programmable Logic Controller

- HMI = Human Machine Interface

- Download = Adding code to the PLC

- Upload = Pulling the program off the PLC

- True = High = 1

- False = Low = 0

- Bool = Single true/ false bit.

- Integer (INT) = 8 bools tied together

- Double Integer (DINT) = 2 INTs tied together

1.3 What is RSLogix 500?

RSLogix 500 is a software program that allows your computer to interface with, and set up, an Allen Bradley SLC 500 or MicroLogix PLC. RSLogix 500 is also the software that enables you to program the PLC and see what is going on in the code. There are four different levels of the RSLogix 500 platform available:

- RSLogix Micro Starter Lite (free)

- RSLogix Micro Starter

- RSLogix 500 Standard

- RSLogix 500 Pro

The main difference between the levels of RSLogix is the type of PLCs that you can control. In the free version you can only program Micro 1000 and Micro 1100 controllers.

RSLogix Micro Starter allows you to control all Micro Logix PLCs. RSLogix 500 allows you to control all Micro and SLC 500 series PLCs. RSLogix 500 Pro gives you the same level of control as the standard version but has additional features. These features are beyond a beginner level and will be covered in another guide.

To fully set up a system for a project, you will need to use additional software. Mainly RSLinx Classic, which is a software program used to establish communication between your program and the PLC, as well as other RSLogix compatible devices.

1.4 What is a PLC?

A PLC is essentially a computer with the specific purpose to control a machine. Most PLCs run on 24VDC, but many can run on 120VAC depending on the specific PLC. The PLC works like any other controller; you give it a set of instructions to perform in a specific order, and the PLC executes your program one rung at a time. In RSLogix 500 you must use ladder logic to write your program. The image below shows a brief example of ladder logic with an XIC instruction (labeled as **SWITCH**) and an OTE instruction (labeled as **LIGHT**). This programming method is called *ladder logic* because it literally looks like a ladder with each specific line of instructions called a *rung*.

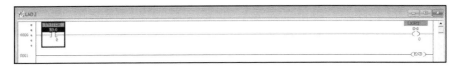

Figure1-1: Simple example of one rung of ladder logic

PLCs also provide physical inputs and outputs to use for controlling machines. We use these physical I/O points to either read data into the PLC or write data to specific outputs. The image below shows a PLC with these physical inputs and outputs. These are what make a PLC extremely useful. A PLC gives you the ability to control specific pieces of equipment based on the inputs going to the PLC, the code we write to handle that data, and the equipment the PLC can control.

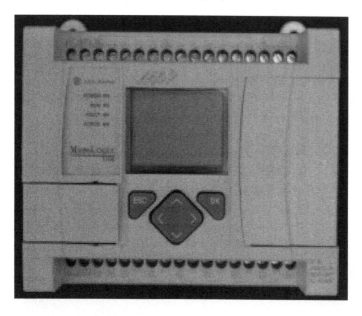

Figure 1-2: MicroLogix 1100 with screw terminal I/O

1.5 Basic Requirements

If you want to use your code to control a machine, you will need at minimum to install RSLinx Classic Lite, which is free software. You will also need a PLC, and a way to connect to that PLC physically. The most common method for connecting to a PLC is via Ethernet, but many older PLCs do not have this capability and require special cables. In a later chapter we will cover how the program knows which PLC to connect to.

If you don't have a PLC but want to practice writing code, you can use a program called *RSLogix Emulate 500*. This program allows you to fully test the logic of your code. These are the basics for what is required to control a simple machine.

In many machines the PLCs default input/output (I/O) space will not be enough, and additional I/O cards will be needed. Luckily most PLCs can expand the amount of I/O ports with additional I/O cards. There are two types of I/O:

- Digital I/O: Inputs and outputs that are either true or false. An input is true when it receives a 24VDC signal and an output is made true internally in the PLC. Output ports typically output the same voltage required by the inputs ports

- Analog I/O: Inputs and outputs that detect varying values. Most analog I/O have a 0-10V or 0-20mA range. If you need to know the temperature, rate of flow, or the pressure in a pipe, you will need to use analog I/O.

1.6 Brief Chapter Overview

Let's look at what you can expect in the upcoming chapters of this book:

Chapter 2 - Simple Programming Principles

This chapter will show you the very basics of what you should know when trying to write any program. This chapter will also cover the basics of programming with ladder logic, and how ladder logic correlates to the PLC inputs and outputs. If you are a beginner, this chapter will help you get a better understanding for how to program, and what is required to control a machine.

Chapter 3 - Interfacing with RSLogix

This chapter will go over working with the actual RSLogix software. What each window looks like and how to navigate through the program. We will also touch base with the RSLinx software in this chapter.

Chapter 4 - Basics of Ladder Logic Programming

Here we take the simple principles from Chapter 2 and put them to work inside RSLogix 500. This chapter will cover XIC, XIO, OTE, ONS, and other basic commands that are required to run a machine. Different PLCs have different capabilities, but these instructions will work for every PLC.

Chapter 5 - Memory Addressing

Here we show you how to assign instructions, such as an XIC, to a static memory location. Unfortunately, RSLogix 500 doesn't have the ability to assign instruction names. This chapter will show you how to navigate through, and use, the memory addressing system.

Chapter 6 - RSLogix Program Instructions

In this chapter you will learn how to find every instruction available in RSLogix 500. You will also learn how to find out what each instruction does, and which PLCs those instructions will work for.

Chapter 7 - Timers, Counters and Integers

This chapter covers the specific timer, counter and integer instructions. You will learn how each one works, and when to use them.

Chapter 8 - Move, Jump and Math Functions

Here we cover moves, jumps, and math functions. This chapter will show you how to set variables in the code, and how to do simple math on those variables. You will also learn how to "jump" to specific sections in your program, because we don't always want the entire program to execute linearly.

Chapter 9 - Peripheral Devices

This chapter will be important if you intend to add additional devices to your control system. A PLC needs to be set up in order to look for something like an HMI or motor controller.

Chapter 10 - Practical Examples

The final chapter will bring everything together by presenting two simple systems to practice with. One will be a simple pump system. The second will be a slightly more complicated bottling system.

2. Simple Programming Principles

If you are new to programming control equipment, or to programming in general, this chapter will show you how to write a program from scratch, as well as how to write it fast and effectively. This chapter will take you through creating the end-goal for your project, which is the first and most important part. After that you will learn about breaking up the project into smaller parts and making a flow chart. Finally, this chapter will cover how to put everything together to see the project as a whole from start to finish, before starting to program.

For the duration of this guide, we will use a simple example of a garage door control PLC. We will walk you through the thought process of creating a program to control a fully functional garage door. The end goal of the garage door will be to open when requested, and not close until there have been no vehicles traveling through the door for over 10 seconds, or if the close-door button has been pushed. We will also consider what will happen when an alarm condition occurs, or the E-stop is hit. Later in this chapter we will look at a diagram of the garage door, along with every input, output and sensor we will have access to for controlling the garage door.

In this chapter we will start by determining the process for how the garage door will work and decide what needs to be in place to control the door. In subsequent chapters we will focus on how to use RSLogix 500. A garage door may seem like a very simple example, but there are multiple aspects to consider in setting up the code for the garage door to work.

2.1 Determine Your Goal

The first step to completing any project is to make sure you understand the end-goal. If you don't first know what the objective of the machine is, then you can't truly finish the project. But don't merely start writing a program when you only have an idea of what the goal for the machine is.

With the garage door example, we know that the door needs to open and close, but we still need to obtain more details before we start working on the project. Does the door open when we push a button on a remote or when we approach the door? When does the door need to close? What if something is stuck underneath the door? Would it be possible to make sure the garage door stays open as long as necessary when something was in the way?

Another important factor is to make sure you know all the data sources that you have access to for your project. If you know what the machine needs to accomplish, but don't know all the signals you will have access to, it will be very difficult to write your program effectively. We need to know if we have sensors that tell us when the garage door is in a certain position, or if the door will only run on a timer. The amount of data we have access to will greatly affect our program in every stage of the project.

In this phase of the project, you should make sure you have all the information you need before starting your program. When working on a project, make sure you ask as many questions as you can in the beginning. Here are a few questions to help get you started:

- What is the project?

- What will the operator of the machine have to do?

- How should the process start?

- How should the process end?

- What order does each phase of the machine have to occur in, or is there an order at all?

- What equipment will be sending information back to the PLC?

- What will the PLC have to move, and what will the operator have to move?

- What should happen in case of a fault or E-stop situation?

- Ask for pictures or drawings of the machine (Since most people are visual thinkers, this can help you visualize the process)

Once you know exactly what your program needs to do, and what information you will have access to, it will be helpful to list the potential problems that could occur. The earlier on in the project you start considering potential problems, the better.

In order to make sure your program runs effectively, you will want to account for situations that could cause the machine to fail and damage the product. This will help streamline your programming process, because you will not have to wait until the program fails during execution to realize certain weak points. Planning for it in advance will save you countless hours of fixing your code down the line.

2.2 Break Down the Process

Have you ever heard the saying "To eat an elephant, you do it one bite at a time"? This is the way in which you will need to approach most programs when you are ready to start writing them. You will need to break everything down into an order of operation and organize it into a logic flow diagram. A logic flow diagram is a chart that shows how the machine will transition from one state to the next, based on certain conditions. An example of a logic flow diagram is shown below.

The flow diagram shown on the next page is very simple, but it already starts to get a little complex. That is why the flow diagram only shows the door open sequence of our garage door example. Breaking the overall process down into small individual processes that are linked can help make writing the program a lot easier. Once the door open sequence has been completed, our program can jump to the door close sequence and run that as its own separate process.

For you to create your program correctly you need to know the exact order each task needs to be completed in, and what state (status of the machine) the equipment must be in before the next task can be completed. For the garage door closing sequence we would want to make sure that the vehicle is out of the way before the garage door can close. This process is the same when designing a program for any piece of equipment, whether it is a full Tesla power train assembly line, or a garage door opener.

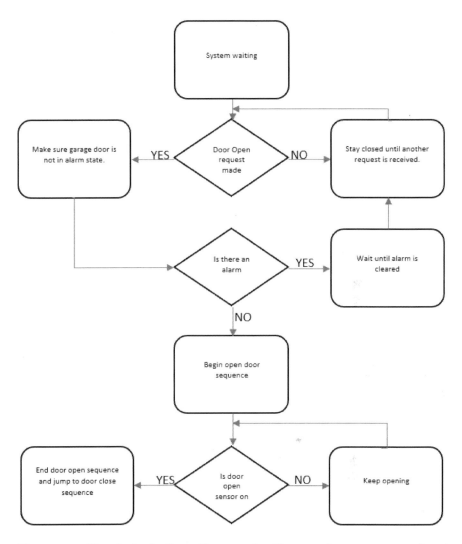

Figure 2-1: Simple logic flow diagram for the opening sequence of our garage door example

One more tip for breaking the project down into smaller parts, is to draw a picture to help you see everything you have to work with, and to determine what physical characteristics you have to work around. This step is not as important as the logic flow diagram, but it can help you move forward with the project if you get stuck. A sketch can help you clearly create the project in your mind. It can help you visualize the machine and each part that needs to move. Then you can start assigning variables from your program to the physical machine in your drawing. The sketch below shows every piece of equipment that we will have to work with for the garage door program.

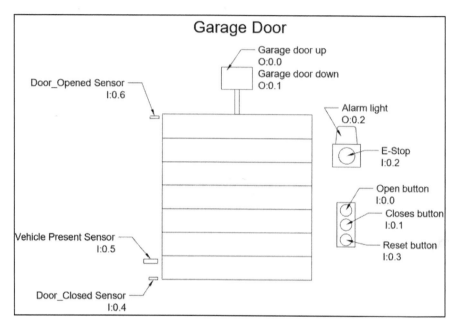

Figure 2-2: Drawing with labeling of our garage door project

After you have laid the foundation for what needs to happen in each part of your program, you can start considering alarms. When you list the potential problems you will come across, you might discover some that cannot be fixed with code. If our garage door starts to close, but something is blocking its path without the sensor picking it up, the door will crash without stopping after the crash. To account for this, we have to create an *alarm* condition that specifies: If the garage door takes longer than 10 seconds to close, while the motor is closing, throw an alarm and stop the motor.

Mechanical failures occur in real life, and your program must account for the possibility of a mechanical failure. Things such as sensors failing, chains breaking, pumps seizing up, and any number of things can go wrong. If your code doesn't double check to make sure everything is the way it should be, or if it doesn't get there in a certain predetermined time, it could cause a lot of damage to the system, the product, or both.

The final aspect to account for before you start programming is the E-stop and recovery. Most (if not all) systems you program in RSLogix 500 will have an emergency stop (E-stop) that is intended to completely stop all motion of the machine. An E-stop isn't just a big red button, it should be an input wired directly to an input on the PLC you are using. Typically, the E-stop signal is always active and turns off when someone hits the button. It is important to plan for how the E-stop could affect the machine overall.

The E-stop will remove power to most of the equipment on the machine when pressed, which is another reason why your program will need to know at which point in the process it is pressed. If someone hits the E-stop and physically moves something, you need to make sure that when the machine starts back up, no damage will occur. This is where the recovery process comes in. You can recover from an E-stop in a few ways: either continue from where the E-stop was pushed, or restart from the beginning of the process. This depends on your specific project and what needs to happen next in the machine process.

If you take the time to build a logic flow diagram and add in possible alarms, an E-stop, and recovery conditions before you start programming, it will save you hours of work altering code one line at a time later on. When looking at our garage door example, we would need to consider what should happen if the garage door is closing and the E-stop is pushed. We could either have it continue closing after the area is cleared or open all the way again and redo the closing process.

2.3 Putting It All Together

If you have gone through all the processes above, then you should have a good idea of how your program should run. An easy place to start writing your program is by creating all the I/O symbol names. For our current example we could start by naming bit I:0.0 to be the *Door_Open* input, bit I:0.1 as the *Door_Close* input, bit I:0.2 as the *E-Stop*, bit I:0.3 as the *recovery button*, and so on for every sensor.

After you have every physical input to your PLC set up in RSLogix 500, and you have your logic diagrams set up, it should be easy to start writing code. Once you start making the program though, you will come across things you didn't account for. This is where iteration comes in. It is still very likely that you will make multiple code changes throughout writing your program. But if you are taking everything into consideration as you make changes, you will be able to finish the program much more effectively.

Programming isn't a quick process most of the time. However, you will probably be able to do a good chunk of the programming rather easily at this point. Creating the base of the program is the easiest part. Adding symbols to all of the I/O you will use, and creating data handling symbols, will help you keep track of everything as you write your program. In RSLogix 500 you can give a symbol name to every single bit, integer, timer, and counter.

16

You will have to make sure that you do your best to name each one of these clearly and appropriately.

For example, we wouldn't want to name I:0.0 = *Top_Push_Button* because this doesn't mean anything to an outsider who doesn't have the machine in front of them. Also to figure out what *Top_Push_Button* does, they will have to dig through the code. If we name the symbol as we did above, *Door_Open*, anyone can immediately tell what that particular symbol is going to be used for.

Make sure to also add comments to the code, to make it easy to follow. Along with symbol names, you can add comments to data points as well. Comments allow for a better description of what each bit is used for, and comments can be added to entire rungs, so you can describe what is happening at each rung. In the image below, you can see an example for how to use symbols and comments in a few simple lines of code. This is a very important practice, because you might have to come back to work on this program in the future. And if you didn't properly create symbol names and comments, it can be daunting to figure out what is going on.

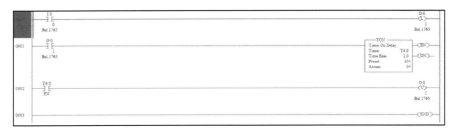

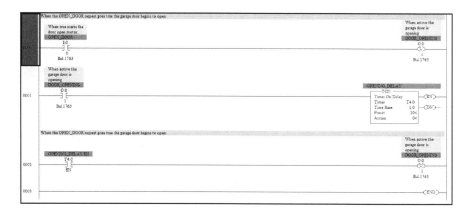

Figure 2-3: The same program without symbols and comments (previous page), and with symbols and comments (above)

In conclusion, remember that the best place to start with a new program is to determine the end goal. Be sure to ask all of the necessary questions when starting your project. After you are confident in what you need to accomplish for your program, you need to break up the project into smaller pieces and figure out the process of the machine. A logic flow diagram is a great tool to help you determine what needs to occur, and in what order it needs to occur for every individual process of the machine.

Finally, you need to be able to see every individual step together as a whole. Create the entire picture in your mind as to what you need to consider when writing your program, and then use comments with good symbol names to help you keep track of the process. You don't need to use all of the tips in this chapter, but if you ever get stuck or find yourself spending hours on a project not making progress, these tools could help you get to your end goal more quickly.

3. Interfacing with RSLogix

In this chapter you will learn the RSLogix interface, and by the end you will know how to navigate through RSLogix with ease. This chapter will also go over the process of creating a new project and opening old ones. You will learn how to organize the main program window, and where you can find every instruction available in RSLogix. In addition to that, you will also learn how to find instructional help with every program.

The first time you open up RSLogix you will see the screen in the image below. All of the navigation windows are the same for each version of RSLogix, with the only difference being the name in the top left corner. Let's get more familiar with the interface.

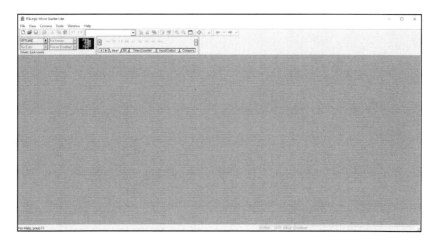

Figure 3-1: Interface panel for the RSLogix software platform

3.1 The Main Header

When you open RSLogix, a project won't pop up right away. You must either start a new project or open an existing one. You can do this by clicking the **File** tab in the upper left-hand corner. This is highlighted in the image below.

Figure 3-2: Tab ribbon for RSLogix with the file tab highlighted

In the file tab you will have three options to choose from. You can either click on the **New** button to start a new project, on the **Open** button to select a previously started project, or you can select a recently opened project from the drop-down menu below the print button. The list that will appear here is from the last three or four projects you worked on. In the image below, you can see that the only recent project we have is the *Garage* file. **Print Setup** isn't very useful until we have a project opened, because we have nothing ready to print at this point.

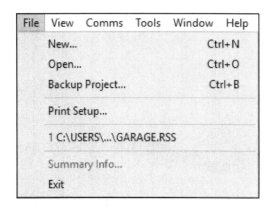

Figure 3-3: Drop-down menu for the File tab

After you have created or opened a project and then click on the **File** button, you will see that the drop-down menu has far more options available. In the image below, you will see all of the additional options that appear.

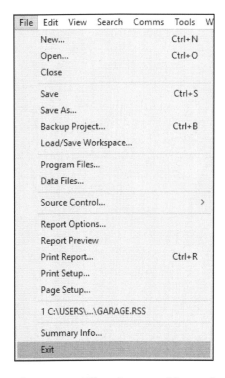

| File | Edit | View | Search | Comms | Tools | W |

New...	Ctrl+N
Open...	Ctrl+O
Close	
Save	Ctrl+S
Save As...	
Backup Project...	Ctrl+B
Load/Save Workspace...	
Program Files...	
Data Files...	
Source Control...	>
Report Options...	
Report Preview	
Print Report...	Ctrl+R
Print Setup...	
Page Setup...	
1 C:\USERS\...\GARAGE.RSS	
Summary Info...	
Exit	

Figure 3-4: File tab menu if a project is open

The **Save** and **Save As** buttons are the same as most programs, and the **Backup Project** button gives you the option to create a backup file for any of your existing projects. If you want to print your project, you have to create a report using the **Print Report** button. Set up the report using the **Report Options** button first. Once you click on the **Report Options** button, the screen in the image below will appear. This screen gives you the option to select how much information you want to print on your report.

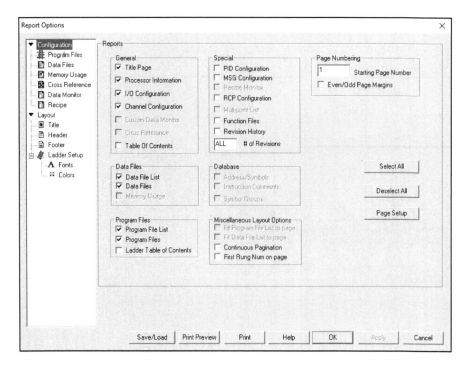

Figure 3-5: Report options pop-up screen

The **Program Files** and **Data Files** buttons pull up all of the program files in your project in one window, or all of the data for the data files of your project in a single window. This book will cover more on the project and data files later on.

The drop-down menu for the **Edit** button is important for using **Verify File** or **Verify Project**. These buttons are used while you are creating your project. The image below shows the edit drop down menu. If you want to add more detail information to your project, you can click on the **Properties** button and a menu will pop-up giving you the option to add detail such as author, keywords, comments, title, subject, and template.

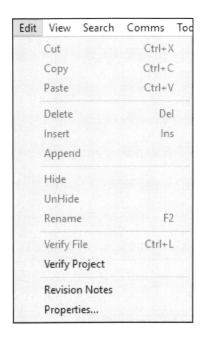

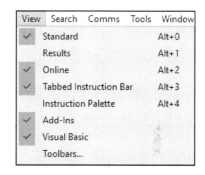

Figure 3-6: Edit drop down menu

Figure 3-7: Drop-down for the view tab

The next tab, the **View** tab, contains a drop-down menu for viewing each available toolbox on the ribbon. The default setup for the ribbon is enough to fully develop your program. The image above is of the RSLogix ribbon with the view tab drop-down. Each of the check marks is a selected item that appears on the header of RSLogix.

The **Search** tab menu has the options for using the **Find**, **Replace**, and **Goto** abilities. These work the same way in RSLogix as they do in any of the Microsoft office programs, note pad, etc. If you use the verify project button that is under the edit tab, and multiple errors are present in your program, you can jump to each one using the **Next Error,** and **Prev Error** buttons in the search tab. The image below shows the drop-down menu for the search tab.

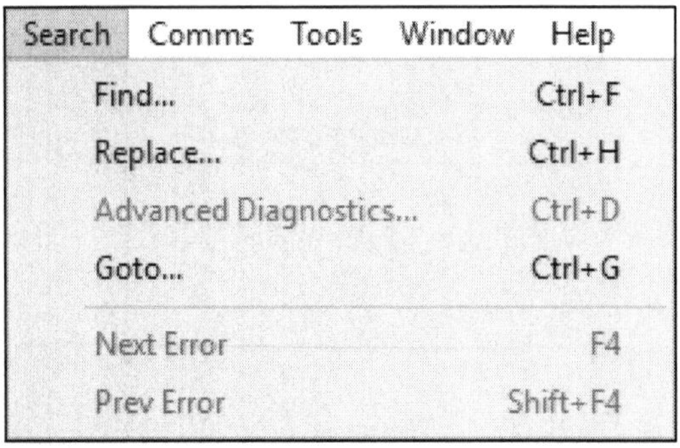

Figure 3-8: Drop-down menu for the search tab

The **Comms** tab is one of the most important tabs in RSLogix, because in the drop-down menu is where you will find the **System Comms** button. This button will pull up the RSLinx application which allows you to match your program up to the PLC. If you do not have RSLinx installed, you will not be able to use this function and won't be able to match up your program to your PLC. In this panel you will be able to see which drivers you have configured.

The comms tab also contains the **Who Active Go Online** button, which starts a scan on all of your current configured networks in RSLinx to try and find a PLC. This guide will go into more detail on RSLinx and how to configure and use drivers in Chapter 9.

The image below shows the comms drop-down menu, and the subsequent image shows the pop-up window for system comms.

Figure 3- 9: Drop-down menu for the comms tab

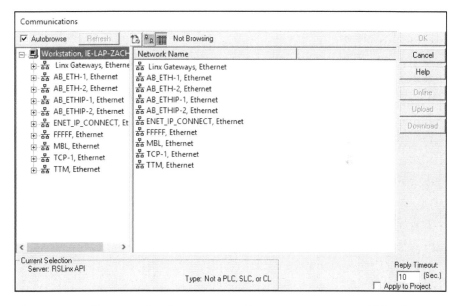

Figure 3-10: Communications pop-up for system comms. This shows a list of currently configured drivers

The **Upload** and **Download** buttons are important in RSLogix as well. To upload in RSLogix means that you are copying the program that is currently running on the PLC to your current open project in RSLogix. When you do this your current project will be overwritten, so make sure you don't upload to a project that doesn't have a back-up file. To Download in RSLogix means to do just the opposite of upload. You copy your current project from RSLogix to the PLC that you are connected to. This process will completely remove anything else currently on the PLC. If you are working with a PLC that is already functioning, then I would recommend first creating a back-up file of the program running on the PLC, before downloading over it.

The **Mode** button is for selecting the mode of the PLC (you have to be actively connected to a PLC to make use of the mode functions). Here you can set the PLC to **Program Mode** which enables you to make changes to the code while online with the PLC (this will be discussed in detail later on), or **Run Mode** which allows the PLC to execute the program currently downloaded to it. You have the ability to do a **Test Single Execution**, which will go through your entire program once to show the state everything is in afterwards. There is also **Test Continuous**, which will keep running through the code until you tell it to stop. This can be useful for testing the logic of your code, but won't be as helpful as actually seeing your program run on the physical machine. **EEPROM** can be used to store or load information if the PLC you are using has this capability, but most of the time this function is not needed.

The **Window** tab gives you an easy way to navigate through RSLogix if you have multiple ladders and data files open. An example of this is shown in the image below. The more windows you have open in RSLogix, the larger the drop-down menu becomes, but the **Arrange** button will always be visible. The arrange button comes in handy when you either have multiple windows open and want to arrange them in a specific order, or if you simply want your project program to appear and take up the full RSLogix window by selecting the **Default Project** mode. This pop-up window is shown in the second image below.

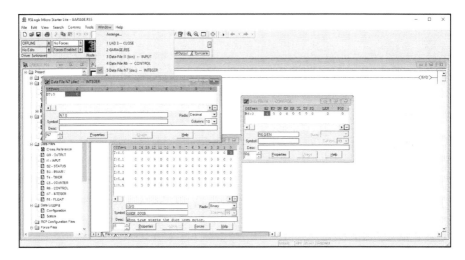

Figure 3-11: RSLogix with multiple programs and data files open

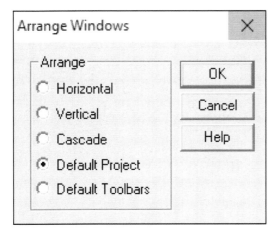

Figure 3-12: This will help you organize the look of your projects in RSLogix

The final tab on the main ribbon in RSLogix is the **Help** tab. This tab is very useful for people that are new to RSLogix, and new to programming in ladder logic in general. The two figures below show the help tab and instruction help pop-up page. **Instruction Help** is where you can see every instruction that is available in RSLogix. Each instruction has a description of what it does, and which PLCs that instruction will work with. If you aren't sure which instruction you should use or how to properly use a certain instruction, this will be a great resource. Not all versions of RSLogix have the same instructional list, because not all PLCs can do the same functions. If the PLC you are using can't use a certain instruction, it will be greyed out in the help guide.

The **Contents** button brings up the entire RSLogix help guide. If you aren't sure about what a certain button is for, or what a certain window is supposed to show you, the help guide will be able to help you find a solution. **Copy Protection** brings up the copy proctection help guide, which explains how files in RSLogix are protected. **User Application Help** is for adding reference matieral to your program if needed.

RSLogix Release Notes will take you to a web page that has all of the release notes for the current version of RSLogix that you are using. **Rockwell Automation on the WEB** gives you a few choices for specific links to click on. Each of these links will take you to a different rockwell website. You can select either **Product Update, Internet Support, Support Library,** or **RSLogix User Forum**. Lastly, **About RSLogix** will open a pop-up screen that will tell you the current version you are using along with a few other informational details about your current version of RSLogix.

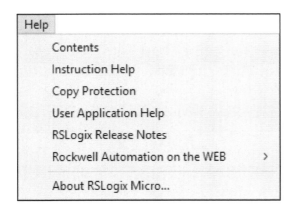

Figure 3-13: Help tab drop-down menu

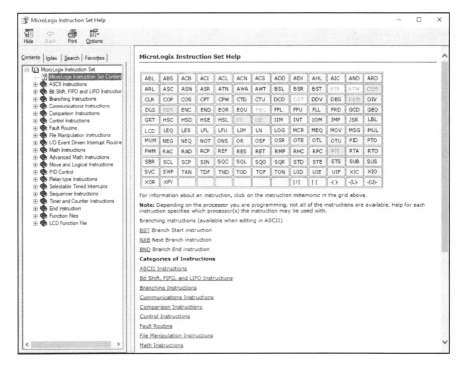

Figure 3-14: The instruction list you have available for your RSLogix software

3.2 The Project Window

Now that you have learned what everything does in the main header, we will move onto the project window as shown in the image below. The project window displays the details for the current project. You will notice several folders under the project folder. We will discuss each one briefly next.

Help is another location to select the same help options discussed in the previous section. The **Controller** folder gives the option to look at controller properties (the controller is the PLC), the status of the processor, I/O configuration, and channel configuration. The **Program Files** folder shows current program files for the project. You can create multiple programs or ladders for each project. The **Data Files** folder allows you to select each of the specific data locations in your project to see what is used and its status.

Data Logging is used for keeping track of certain variables. The **RCP Configuration Files** are only used with the MicroLogix 1 series C processor and is for saving custom lists of addresses. We will not be covering this option in this guide, since it's not vital to using RSLogix. **Force Files** are used for temporarily forcing certain bits either on or off. Custom data monitors, custom graphical monitors, recipe monitors, and trends are all forms of data logging that can be set up. Finally, the **Database** folder is for viewing comments, symbol names, and other descriptions you have made for your code.

To avoid being overwhelmed, we will cover the most important items for a beginner from the project window, instead of going through each line in detail at this stage.

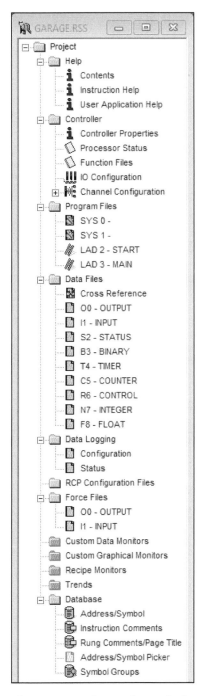

Figure 3-15: The project window

Under the **Controller** folder, the **Controller Properties** button will open a pop-up to display your controller setup. In this window, the **General** tab is the most useful. In this section you can select the processor type and processor name, and you can see the amount of memory your program has left.

The **IO Configuration** button is used to add more digital and analog I/O. The image on the next page shows the I/O configuration pop-up. Depending on the type of PLC you are using, there will be different options of I/O cards that you can add to your project. In the image below you can see that there is one digital input card with 8 slots ranging from 10-30VDC, and one digital output card with 8 slots added to the PLC.

These are just a few of the options available. To add an I/O card to your current project, click on the I/O card you want to use and drag it to the correct slot number.

It is important to make sure that the I/O card you select in the window matches up exactly to the I/O card that you are physically using on the PLC. If it doesn't match, it won't work. It's also important to mention that different PLCs will have a different number of I/O cards they support. Some PLCs don't have any expansion capability.

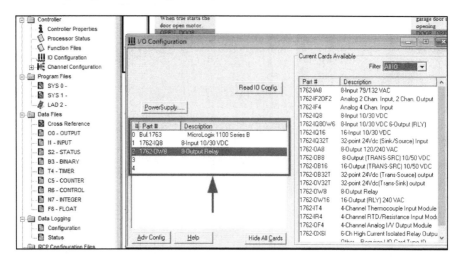

Figure 3-16: The I/O configuration for a MicroLogix 1100 B. It only has the ability to add 4 I/O cards

The **Program Files** folder will be discussed in greater detail in the following chapters, because it is the primary folder we will use when writing our program. The **Data Files** folder is very useful for helping to keep track of all the symbols used in the program. This folder will also be covered in more detail in the following chapters.

Data Logging, RCP Configuration Files, Custom Data Monitors, Custom Graphical Monitors, Recipe Monitors, and **Trends** are all used for monitoring data in different ways and do not work for every PLC available. Because of this, we don't go into more detail on any of these choices for now. The **Force Files** are used to quickly detect if you have any bits currently forced on or off. To force a bit on or off means to make it true regardless of the current state of the program. You can also force bits on and off inside the pop-up windows, an example of this is shown below.

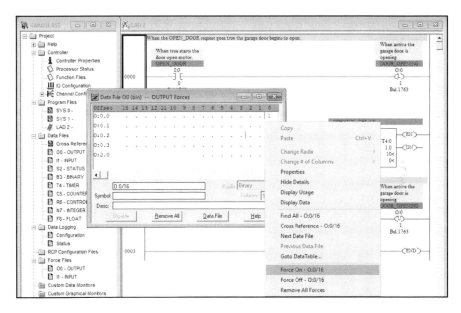

Figure 3-17: Bit O:0/0 forced ON, with every other bit in its normal state

The final folder is the **Database** folder. In this folder you will be able to see everything relating to comments and symbols. The **Address/Symbol** button pulls up a list of everything that has a symbol and specifies what that symbol is and what its address is. **Instruction Comments** pulls up a window that allows you to see the comments you have made, and to search for comments you made at a certain bit address. **Rung Comments/Page Title** pulls up comments that you made to describe an entire rung, and their location.

Symbol Groups allows you to create symbol groups which help to manage code if your program starts to get too large.

3.3 The Quick Access Toolbar

The final aspect of the interface we will cover is the quick access links, shown in the image below. In this toolbar you will find the most commonly used instructions separated into categories. The **User** category has all of the most common bits used when first creating a program. In this tab you can add a new rung, create a parallel instruction, add your normally open/ closed bits, as well as the output bits. This will all also be discussed in more detail in the following chapters.

Next is the **Bit** tab which shows the same bits as the user tab, but with the ONS and a few other options. The next tab you will see is the **Timer/Counter** tab. This tab is for quick access to the TON, TOF, and Count instructions. There are a few other quick instruction tabs you can select, but these are the ones you will use most often when writing your programs.

Figure 3-18: The box highlights the quick access toolbar. User is the default and most often used list.

In conclusion, there are many functions in RSLogix and it will take time and practice to get to know them all. If you get stuck on what type of instruction to use, or what a certain button, screen, or folder is for, RSLogix has a very in-depth help guide for any interface information you might need that was not covered in this chapter.

34

If you are ever uncertain about the status of your project or program, you can refer to the data files section of the project window to see the status of your project. A large amount of the detail pertaining to the interface will be covered in the remainder of this book, which is why certain areas may have been glossed over in this chapter.

Thumbs up or thumbs down?

I would love your feedback on the content and format of this guide, good or bad. I use your input to add more value to revisions of this guide and future guides. So let me know what you liked, and what you didn't like so far, through a short review on Amazon. My wife and I read each and every one of them.

a

---

4. Basics of Ladder Logic Programming

In this chapter you will start to learn how to program on the RSLogix platform. We will start by covering simple logic such as XIC (normally-open), XIO (normally-closed), OTE (normal output), OTL (latched output), and OTU (unlatched output). After going over the basic instructions we will discuss methods, common practices guidelines, and simple mistakes to avoid when starting a program. Finally, we will cover basic tools you can use to test your program. Data addressing will be mentioned a lot in this chapter, but won't be covered in detail until Chapter 5.

4.1 What is Ladder Logic?

So far, we have shown a few examples of ladder logic, but it hasn't been discussed in detail. Ladder logic looks very similar to circuit diagrams of relay logic. This was done to make the programs easy for electricians and maintenance workers to read and create, who don't necessarily have prior programming experience. Because of this, you can think of your program layout as a type of circuit.

Once your project is created, click on the *Program Files* folder on the left side of your screen and select the *LAD 2* file. This is the starting ladder program shown in the image below.

To start adding instructions to the new program, click the instructions in the top highlighted box in the image and drag them to where you want the instruction to take place.

The inserted instructions are read from left to right. The left rung is the starting place or *Power Rung*, and the right rung is the ending or *Common Rung*. The instructions you place between the power and common rungs are your actual program. The second image below shows the flow of a ladder program in RSLogix.

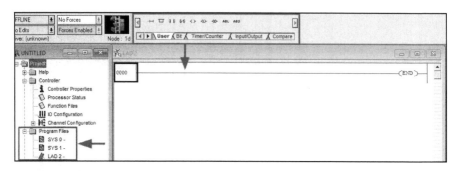

Figure 4-1: Click or drag any of the items in the highlighted bar onto the desired rungs

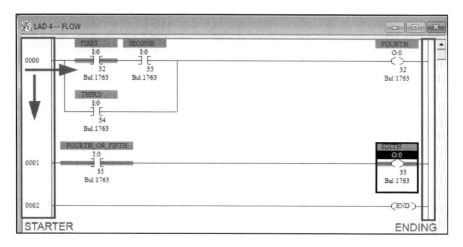

Figure 4-2: Arrows indicate the order of instructions. The horizontal arrow direction is first, and vertical arrow direction is second

Starting from the top left, the program is read first by searching linearly across to the right, before reading the next line below it. Just like you are reading this paragraph.

The program reads through the test instructions before energizing the outputs. A test instruction refers to an XIC, XIO and several other instructions, which have to be true in order for the program to keep reading a certain rung. Rungs don't have to only be one linear set of instructions either. You can use something called a parallel rung to test multiple instructions simultaneously. The image below shows in more detail how a program reads. If an XIC or XIO instruction is true, it will have green markings around it. The same goes for OTE instructions.

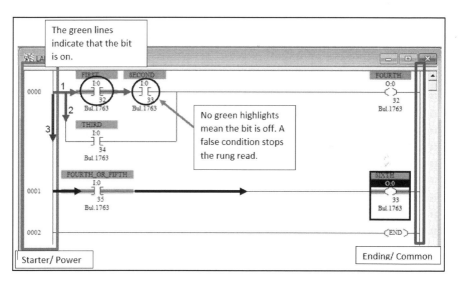

Figure 4-3: The arrows indicate the order the program is read

The reason we focus on the flow of the program in this much detail, is because every instruction isn't read at the same time. You can use the program scan sequence to your advantage when controlling certain systems.

If you are having trouble getting your program to work, or if you need to make sure a certain instruction is true before the next step can happen, you can use the way RSLogix reads through the program to your advantage.

4.2 XIC and XIO Instructions

The XIC, or Examine If Closed as it is called in RSLogix, has two states being either on or off, similar to a switch. A simple example of an XIC instruction would be a switch wired into input I:0/0 on our PLC. If we flip the switch on (i.e. closed) the XIC assigned to data location I:0/0 would become true. When the switch is off the XIC for I:0/0 would become false. The opposite is true for XIO (Examine If Open) instructions. When the switch is turned off it will read true.

To give a better idea of how to use these instructions, we will go over a simple example. If we wanted to use a PLC to control lights for a room and we had two light switches, how could we use the XIC and XIO bits to control the lights? First we need to identify exactly what we want to happen. For this example, we want each switch to be able to turn the lights either on or off at any time, no matter the state of the other switch.

We also need to identify a starting state, possibly both light switches start in the off position and the lights are off. From here we can start thinking about how to program the switches. For this problem we can set up a simple bit table to make sure we have the correct sequence figured out. In the table below (just like for the PLC) a switch in the on state is equal to 1 and in the off state is equal to 0.

Switch 1	Switch 2	Lights
0	0	0
0	1	1
1	0	1
1	1	0

Now that we have identified the goal, the starting requirements, and what will cause the lights to turn on, we can start programing. Since either switch 1 or switch 2 has to be in the on position for the lights to turn on, we will need to use at least one XIC instruction for each switch. And since we know the lights need to turn off when both switches are in the on position, we will need to use XIO instructions to check if both switches are on. The image below shows one solution to our light switch example.

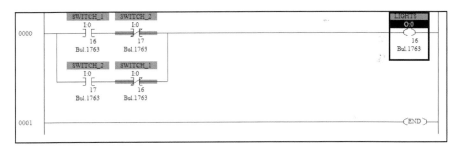

Figure 4-4: XIC instructions are on the left with switch 1 at the top. XIO instructions are on the right with switch 2 at the top

4.3 OTE, OTL and OTU Instructions

For our example to work with a PLC, we need output instructions to turn on the lights. This is where the OTE or Output Energize instruction comes in.

The way this instruction works, is that when every instruction placed prior to the OTE is activated, then the OTE will be made true. It is possible to have a rung where the only instruction is an OTE, which means that the OTE bit would always be true. In our example O:0/16 is the data location used to indicate that the lights have been turned on. The OTE is only active while every instruction in-front of it remains true.

An OTL instruction is basically the same as an OTE, with one very important difference. An OTL is an Output Latch bit, which means that once it becomes true the OTL will stay true even if the instructions before it on the rung become false. It is a maintained instruction. Turning to our example, what would happen if we used an OTL instead of an OTE? Once we turned both light switches off, the lights would remain on.

The only way to de-energize a data location that is assigned to an OTL bit is to use an OTU (Output Unlatch) instruction at the same address. This means that if the data address O:0/16 was used as an OTL, we would need a second location where O:0/16 is used as an OTU to turn that data location off. Using the latch and un-latch instructions we can program the light switches in a different way as shown below.

The reason why latches are so important is because you can use them in multiple places for the same data address. The way RSLogix executes its program causes issues when there is conflicting data. For example, in the second image below, if two OTE instructions were used instead of the latch unlatch method it is possible to have a conflict at data address O:0/16. The conflict would occur if switch 1 is on and switch 2 is off, or vice versa. When this conflict occurs, RSLogix executes the instruction that comes last.

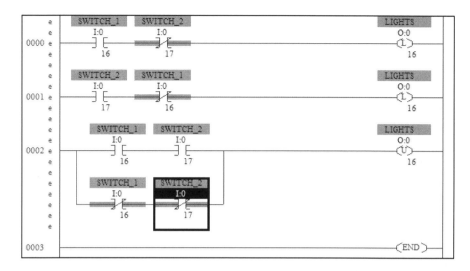

Figure 4-5: We accomplish the same goal to turn the lights on and off using OTL (shown towards the right with an L) and OTU (shown towards the right with a U)

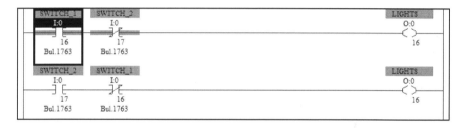

Figure 4-6: Conflicting data using two OTE instructions

4.4 Basic Tools and Setup

So far, we have covered how to use and insert XIC, XIO, and OTE instructions, but we have yet to cover how to assign meaning to these instructions. After you add in a new instruction you have two choices when assigning that instruction to a specific data location. You can either type in the exact data location, or you can type in the symbol name that has already been assigned to a data location. Once you specify a symbol for a specific data location, it is there permanently.

The two images below show two examples of how to assign a data location to an instruction. In the first image you can double click on the newly added instruction type in the desired data location. If that data location has not been assigned a symbol name yet, the box on the right will pop up allowing you to add a symbol name and comment. The second image shows the data file for I1 inputs, with the specific data address designated as SWITCH_1. It is important to note that every instruction used in this chapter is a single bit out of a word.

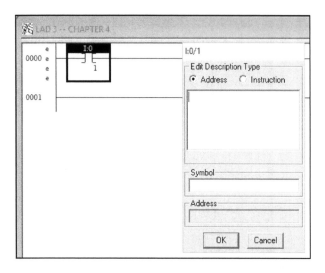

Figure 4-7: Assigning a new instruction to a data location

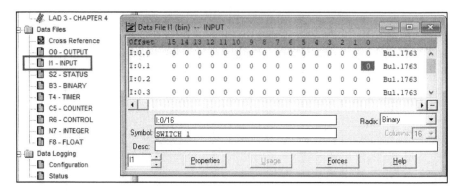

Figure 4-8: Alternate method of assigning a new instruction to a data location

One practice that can really help you speed up your programming, is to go through all of the data locations that you know you will need in advance and assign them symbol names. When you add in a new instruction it is easier and quicker to write in a symbol name than a data location. Another helpful tip is that every rung you use in RSLogix has to have an output, but they don't need any other prior instruction.

As you are building your program you will notice little "e" symbols next to each rung on the left-hand side as seen in the image below. These little symbols are to indicate that the rung either has an error or has yet to be verified. Verifying is a function in RSLogix that tests to make sure there are no errors.

An example of an error would be an instruction without an assigned data location, or a rung that doesn't have an output at the end of it. If a rung does have an error, RSLogix has an error message bar at the bottom of the screen that will pop-up and tell you what is missing. The subsequent image below shows an example of an error message. To verify a rung, you can right-click over the "e" symbols and select *verify rung*. If you want to verify the whole program or project, you can click the buttons circled in the image.

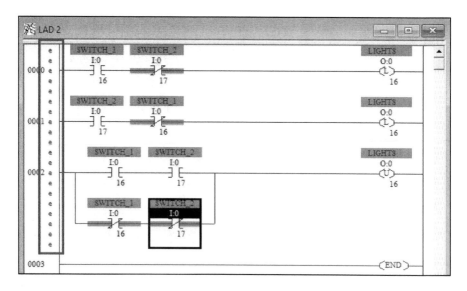

Figure 4-9: e Symbols indicate errors or unverified rungs

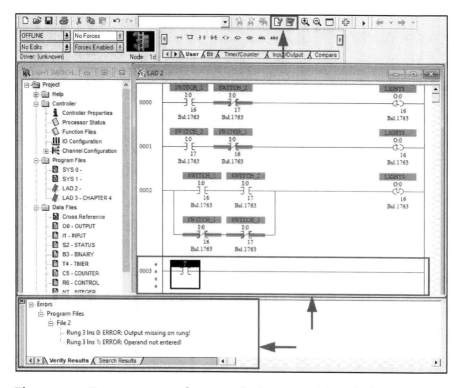

Figure 4-10: Error message shown at the bottom of the window

5. Memory Addressing

This chapter will cover each of the data files in detail. We have already touched on the input and output files, but they will be covered in greater detail here. The data files will be discussed in sequential order from output O0 to Float F8. After we've discussed each data file, we will go over a few tips on how to use these data files to your advantage.

5.1 Outputs O0 Data File

The data in the output data file is used for physically powering outputs. The size of the output data file varies depending on the type of PLC you are using, and on the number of expansion modules used. When using an output data location, if it starts at O:0/X the 0 indicates that the physical outputs are located directly on the PLC, but if the output starts at O:1/X the physical outputs are located on the first expansion module. For example, a light wired to O:3/7 means that it is wired to the third expansion module in position 7.

5.2 Inputs I1 Data File

The inputs data file is very similar to the output data file, in that they are physical locations that you will wire to. The expansion modules work the same for both inputs and outputs in the way you reference different modules. There are two types of inputs: *Analog* and *Digital*.

A digital input is a single bit input indicating one on/off position. An analog input is a byte worth of data, and the degree of accuracy depends on the module used. To use the data location for an analog input, write I:3.0 or I:3.1. The image below shows an input data file with an additional 8 port digital input module at channel 1, and 7 analog input ports in the module 3 position.

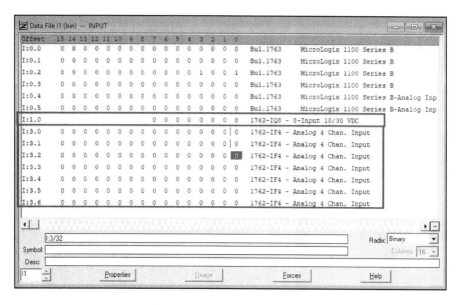

Figure 5-1: An input data file that has expanded due to additional modules attached to the PLC

5.3 Status S2 Data File

The status data file doesn't provide a data location to use for programming like the previous data files discussed. The status file provides information on the controller's current state. This data file is very useful for troubleshooting problems with your RSLogix project. It can help you determine if the problem is with the program or with the controller. There are ten different categories of information you can find in the S2 data file.

48

The following few paragraphs will cover the status data file in more detail, but the information is not necessarily needed for creating most projects. It can be useful if you are having communication or mathematical troubles in your program. The image below shows the pop-up box that appears when you double click on the S2-STATUS data file.

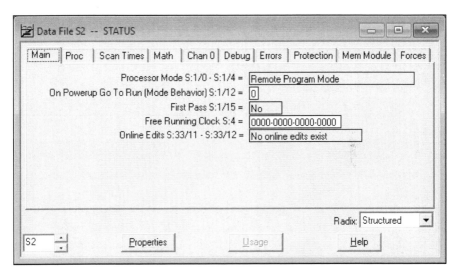

Figure 5-2: S2 data file pop-up menu

Main is the first tab and shows the current processor mode (remote program mode, run mode, test single, or test continuous). It also shows whether the PLC will go into run mode at start up, if there are any current online edits, if there is a first pass scan set up, and the free running clock. **Proc** displays the information of the current PLC the program is meant for. When starting your new project, you select the PLC type and your program can only work for that PLC type, unless you convert your program. If this is needed the RSLogix will usually prompt you whether you want to make this change.

The **Scan Times** tab keeps track for how long it takes the PLC to complete one full scan of the program, usually it is in the micro to low millisecond range. Of course, this number grows the larger and more complex your program becomes. The **Math** tab is used for dealing with large number addition, subtraction, multiplication, and division. There are nine different pieces of data in the math information tab, which won't necessarily need to be used unless you are manipulating large numbers.

- *Math Overflow Selected S:2/14* has to be set to 1 for 32-bit addition and subtraction, otherwise you can only do 16-bit addition and subtraction.

- *Overflow Trap S:5/0* is set to 1 by the controller when an overflow has occurred during a math instruction.

- *Carry S:0/0* is set to 1 when a math carry or borrow occurs.

- *Overflow S:0/1* is set to 1 when the results of a math operation are too big to fit in the destination.

- *Zero Bit S:0/2* is set to 1 when the results of a math operation equal zero.

- *Sign Bit S:0/3* is set to 1 when the results of a math instruction are negative.

- *Math Register S:13-14* has registers used to produce 32-bit signed math operations.

Chan 0 is the channel 0 communication status for the MicroLogix 1100/ 1200/ 1400/ 1500.

- *Processor Mode S:1/0-1/4* indicates the status of the processor, same as in the main tab.

- *Node Address S:15 (low byte)* is the current node address of your processor.

- *Baud Rate S:15 (high byte)* is the current communication speed of the controller.

- *Channel Mode S:33/3* equals zero when the communication port is in user mode, and when set to 1 the channel 0 is in system mode.

- *Comms Active S:33/4* is set to 1 when the controller receives valid data from the RS-232 channel. After 10 seconds of no data this is set back to 0.

- *Incoming Cmd Pending S:33/0* is 1 when the processor detected another node on the channel 0 network which requested information. It is set back to 0 when the processor serviced the request.

- *MSG Reply Pending S:33/1* is set to 1 when another node on the channel 0 network supplied the information the processor requested. This bit is reset when the processor stores the information.

- *Outgoing MSG CMD Pending S:33/2* is 1 when one or more channel 0 messages in your program are enabled and waiting, but no information is being sent out.

The **Debug tab** helps determine if the processor in the PLC is in suspend idle mode. The suspend code is not equal to zero when the PLC is in suspend idle mode. While the suspend file contains the program file number in which a true suspended instruction is located. The **Errors** tab shows the default errors that are present in the processor.

- When *Fault Override at Powerup S:1/8* is set to 1 it will cause the major error and minor error bits to reset on power up, if the processor had previously been in the REM run mode and had faulted. The controller then attempts to go back into REM run mode during startup. This bit must be set offline.

- When *Startup Protection Fault S:1/9* is set and power is cycled while the controller is in REM run mode, the controller will execute the user fault routine before running through the first scan of the program.

- *Major Error Halt S:1/13* becomes true when a major fault occurs.

- *Battery Low S:5/11* is set when the PLC battery is low (if it has a battery).

- *ASCII String Manipulation Error S:5/15* is set to 1 when processing an ASCII string instruction that is greater than 82 characters.

- *Fault Routine S:29* is used to set a program file number to be used in all recoverable and non-recoverable major errors. You can then program the ladder logic of your fault routine in the file you specify.

- *Major Error S:6* is where a hexadecimal code is entered by the controller when a major error is declared.

Protection helps to protect certain information in your project. You can deny future access by prohibiting online monitoring of the processor program, unless you have a matching copy of the project. The lost overwrite protection will indicate when protected files are overwritten. The **Mem Module tab** is for memory modules that might be attached to your PLC processor.

- The processor sets *Memory module loaded on Bool S:5/8* to 1 when a memory module program has been transferred to the processor.

- *Password Mismatch S:5/9* is used to inform your application program that an auto loading memory module is installed, but not loaded due to a password mismatch.

- *Load Memory Module on Memory Error S:1/10* is used to transfer a memory module program to the processor if a processor memory error is detected at power up.

- Setting *Load Memory Module Always S:1/11* allows you to overwrite a processor program with a memory module program by cycling power.

- Setting *On Powerup Go To Run (Mode Behavior) S:1/12* to 1 places the processor in run mode on power up.

- *Program Compare S:2/9* causes no modification of the program on the memory module in run mode.

- *Data File Overwrite Protection Lost S:36/10* is used to determine if retentive data following a memory module transfer is valid.

The final tab, **Forces,** helps you determine if there are any active forces. The forces enabled line is set to enable forces on the controller, while the forces installed line indicates whether forces are active.

5.4 Binary B3 Data File

This data file has not been discussed yet, but it is very useful for programming. This file is very similar to the input and output data files, except these bits are only used in the internal memory. This means that no physical inputs and outputs control, or are controlled by, these bits. To use the binary data location, you will have to write the address as B3:X/Y. This data file is very useful because it gives you the ability to process input data, use and manipulate it, and then send the outcome to the desired output locations. The file size for B3 depends on the PLC version. The higher the version, the larger the data file will be. B3 is only intended to be used for individual bits, but can be used as an integer if need be.

5.5 Timer T4 Data File

As the heading suggests, the T4 data file contains the timers. You can use timers for delays to turn certain bits on or off. The timer instruction and data file are different than any of the instructions we have covered so far. The instructions for timers are either TON (Time On Delay) or TOF (Timer Off Delay), more on these instructions will be covered later. The timer data file is capable of more than just a countdown timer to delay an action from happening, each timer also has several bits that go with it. Along with the actual T4:0 timer there is a T4:0/EN, T4:0/DN, T4:0/PRE, T4:0/ACC, and T4:0/TT. Each of these bytes/bits gives you a different capability. The image below shows the T4 data file.

T4:0/PRE is the preset amount of time that you want the timer to run for. T4:0/ACC and T4:0/TT keep track of the amount of time that has passed by since the timer went active. T4:0/EN is the bit that turns the timer on and starts accumulating time. T4:0/DN is the timer done bit. This bit only goes active if the T4:0/EN is active, and the timer has finished.

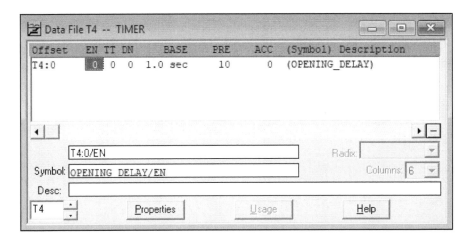

Figure 5-3: T4:0 current state. This timer is currently set to the sec time base

Just like every other data file, the more powerful the PLC you are using, the more timers you will have access to. One other aspect that is different with the timer data file is that you don't name each individual bit. You can only name the full timer. From the image above, you can see that the symbol name is OPENING_DELAY. For this specific timer, if you wanted to call the done bit, you would have to give an XIC or XIO instruction the OPENING_DELAY/DN symbol name.

5.6 Counter C5 Data File

The counter data file is very similar to the timer data file, in that each counter has multiple bytes and bits of data to use. Counters will be covered in more detail in the next chapter, but the fundamental use of a timer is to count the number of times something turns on or off.

C5:0/PRE is the preset number of counts you set before the counter is done. C5:0/ACC shows the current number of counts that have occurred so far. C5:0/CU is the bit that causes the accumulated value to increase when it is turned on.

C5:0/CD is the bit that increments down when it goes active. C5:0/DN is the counter done bit that becomes positive when the preset and accumulator values are equal. When the counter is done you have two options. You can either set up another counter to subtract down to get back to zero, or you can use a reset instruction which will be discussed in the following chapter.

5.7 Control R6 Data File

This data file is used primarily for storing data. The data stored here is the length (length of data), pointer position (where the data has to go to), and status bits for a specific instruction. The instructions are shift registers and sequencers. These instructions will be discussed in more detail later on, but a brief example of a shift register would be something like BSL (Bit Shift Left). Every time this instruction is executed, the data loaded into that instruction shifts the pattern of data through the array to the left. The table below shows the data going through this process, with the image that follows showing an example of how to use this instruction.

	Bit 7	Bit 6	Bit 5	Bit 4	Bit 3	Bit 2	Bit 1	Bit 0
Original	0	1	1	0	0	1	0	1
Shifted	1	1	0	0	1	0	1	0

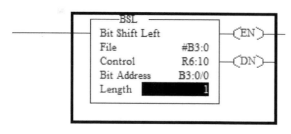

Figure 5-4: Bit shift left instruction.

56

5.8 Integer N7 Data File

This data file is where most of your bytes or numbers will be located. Each integer consists of 16 separate bits, 0 through 15. The integer data file gives you the ability to manipulate analog input and output data, to do math instructions (such as add, subtract, multiply and divide), and to handle register data. A single integer can handle a number from -32768 to 32767. This wide range is very useful when keeping track of input and output analog data. It also gives you far more accuracy when monitoring systems. An analog input takes a 0-10VDC signal and converts it into an integer. Typically, you would set up the PLC so that a 0VDC input equals zero for the integer, and a 10VDC input would equal 32,767.

5.9 Float F8 Data File

A float is a number between 1.1754944e-38 and 3.40282347e+38. A float is very similar to an integer, except you can have decimal point values and it is much larger. Another key difference with the float is that you cannot manipulate the data as much as you can with an integer. You can only look at the decimal value of the float and are not able to expand it out into binary bits. This means that a float data type can only be used with instructions that move, compare or manipulate values.

5.10 Data File Tips

You probably won't need all of these data files in every program you write, but it is important to remember that they are available. There are also different ways you can use some of these data files. Something we would recommend doing would be to use each input bit to turn on a bit in the B3 data file. Using our light switch example, switch 1 was linked to

I:0/16. However, we would recommend using switch 1 to activate B3:0/0, and then use B3:0/0 to active the light output. This way if a physical input breaks down, we don't need to write large portions of code to rectify it, we only have to change which input powers B3:0/0. This will be shown in chapter 10 when going through an in-depth example.

If you have been playing around with the data files, you might have noticed that B3 and N7 can be used in different ways. Even though B3 is designated as a binary, it can be used as an integer because B3:0/0 through B3:0/15 can be combined into B3:0. The way this works is that the PLC will convert a decimal input into binary bits. If you move the decimal value 11 into data address B3:0, the bits B3:0/0, B3:0/1 and B3:0/3 will become true. The table below shows the conversion of the number 11 from decimal to binary.

B3:0	Binary Value
B3:0/15	0
B3:0/14	0
B3:0/13	0
B3:0/12	0
B3:0/11	0
B3:0/10	0
B3:0/9	0
B3:0/8	0
B3:0/7	0

B3:0	Binary Value
B3:0/6	0
B3:0/5	0
B3:0/4	0
B3:0/3	1
B3:0/2	0
B3:0/1	1
B3:0/0	1

Perhaps the greatest use of data files is that you can see the state of large sections of data all at the same time. It is possible to have every data file open at the same time, to see the state of each bit, the value of each INT and FLOAT, and to change the state or value of each piece of data.

RSLogix allows you to force bits on or off, which come in very handy when testing your program without it actually being fully hooked up. If you have an input, such as a proximity sensor, and want to double check that the program reacts in the correct way to the sensor being on, you can force it on. A force remains active until you click on the same bit and remove the force. The image below shows a few ways to activate a force. If something isn't working quite right, it is always helpful to do a quick check for any forces that might have accidently been left on.

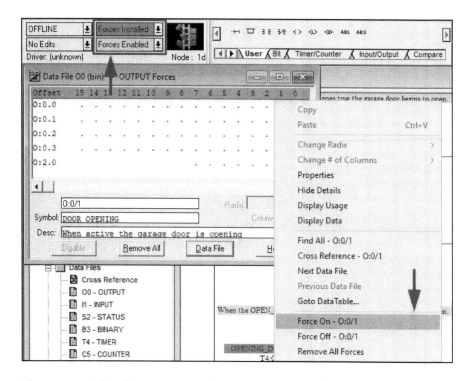

Figure 5-5: If a force was active at any output it would display a 1 instead of a dot. Right click on a bit to open up the option to install a force.

In addition to forcing bits on or off, RSLogix gives you the ability to simply toggle a bit on or off. If we were to right click on switch 1 from our light switch example, we can toggle it on or off. You will see the instruction change state from on to off when selecting the toggle. A toggle is different from a force in a couple of key ways. First, if the PLC is powered down the toggled bit will go back to its original state. Second, a toggle can be overwritten by the program. This means that if you toggle a certain bit off, but that bit has either instructions or a physical input telling the PLC that it should be active, the bit will turn back on.

A good time to use this toggle feature would be to test a sensor, or to turn an OTL bit on to see if the rest of the program runs as planned and unlatches at the end.

It's also possible to create a new data file for any of the data types we discussed, with the limiting factor being the processor of the PLC you are using. The numbering scheme continues for each new data file. If you make a new integer data file, for example, it will be labeled N9. You can also assign a more descriptive name, such as B10 LIGHT_SWITCHES if you wished. To create a new data file, right click on the data files folder and select the *New* button. When you do this, you will see a new pop-up appear as shown below. Here you will be able to select a new data type field. There are also more choices than the original 8 starting data type fields. You will have the additional options of a string, long, message, PID and programmable limit switch.

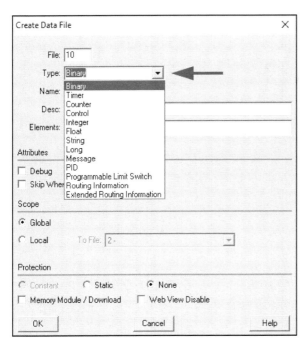

Figure 5-6: The different data type files available to create

Going over the additional data types briefly, the string (ST) data type allows you to store string values. The long (L) data type is a 64-bit complete integer, which allows for the use of large numbers. The message (MG) data file is used to send communication commands to peripheral devices, which can't be directly controlled by the PLC. The PID (PD) data type is used to store the parameters for PID instructions. This instruction will be covered in more detail in the next chapter. The remaining data types will require additional modules for the PLC and are not very likely to be needed for most applications, so they will not be covered in this book.

In conclusion, each data file is useful in its own way for organizing and manipulating data. In addition, you aren't limited to only the initial data files that the project starts with. The biggest lesson to take away from this chapter is that there are multiple ways to manipulate data, and multiple ways to interpret data. Opening a data file shows you the current state and value of each bit or byte for that data file. In the following chapters, the examples will get more in-depth and cover more of the capabilities of the data files.

6. RSLogix Program Instructions

This chapter will provide you with a brief overview of most of the instructions available in RSLogix. Instructions that you are more likely to use will be covered more in-depth, but by the end of this chapter you should have a pretty good idea of what you can use each instruction for. We will also show you how to find all of the information you need about a given instruction, such as what PLCs can be used, what data types you need, and if there are any additional instructions associated with it.

Since we have already gone over the XIC, XIO, OTE, OTL, and OTU instructions they will not be covered in this chapter. Likewise, the CTD, CTU, RTO, TON, TOF, MOV, LBL, JMP, JSR, INT, and SBR instructions will be covered in detail in the following chapters, therefore they will not be covered in this chapter. The math functions will still be covered in this chapter, because not all of them will be covered in Chapter 8.

There is a list in RSLogix that covers every instruction that you have access to when writing a program. Click on the **Help** tab in the main header to open the drop-down menu. From the drop-down menu select **Instruction Help**. The image below shows the drop-down menu and instructional help guide. The instruction list will have certain instructions greyed out, depending on your version of RSLogix and the PLC you are using. If it is greyed out it means that your PLC is incapable of executing that instruction. Many of the instructions that deal with data handling for communication and mask protocols won't be covered in detail in this book. This is because communication is a very broad topic and it changes depending on the PLC and your desired device. This is a subject that will be covered in detail in another guide.

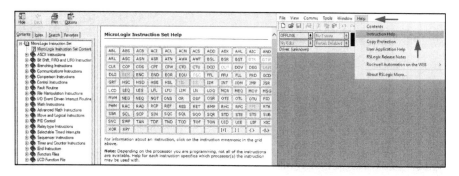

Figure 6-1: Instruction help guide pop-up window

Starting from the top-left corner with the **ABL** instruction. This is the ASCII Test Buffer for Line. This instruction is used to determine the total number of characters in the input buffer. An example of the instruction is shown in the image below. The channel number represents the RS-232 port (channel 0 in this case). The control field is where you will store the control register. The characters field shows the number of characters in the buffer. When the input for this instruction goes from false to true, the enable bit is set and the instruction is put in the ASCII queue. After the instruction is finished, the RN bit and DN bits are set and the number of characters are set in the control register. If the number of characters is greater than 0 the FD (Found) bit is set high.

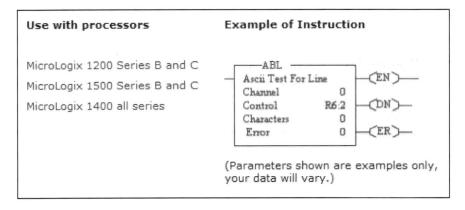

Use with processors	Example of Instruction

MicroLogix 1200 Series B and C

MicroLogix 1500 Series B and C

MicroLogix 1400 all series

```
    ┌──ABL ──────┐
  ──┤ Ascii Test For Line │──(EN)──
    │ Channel        0 │
    │ Control      R6:2 │──(DN)──
    │ Characters     0 │
    │ Error          0 │──(ER)──
    └────────────────────┘
```

(Parameters shown are examples only,
your data will vary.)

Figure 6-2: The ABL instruction

The **ABS**, or Absolute, instruction is a math instruction. It simply outputs the absolute value of a number that you send to the source parameter. After the instruction is executed, the value is placed into the destination parameter and goes to your desired destination. The image below shows the ABS instruction.

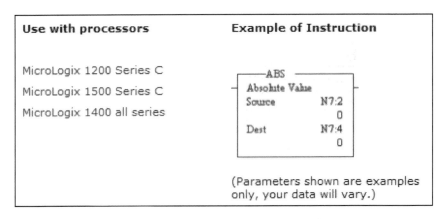

Use with processors	Example of Instruction

MicroLogix 1200 Series C

MicroLogix 1500 Series C

MicroLogix 1400 all series

```
    ┌──ABS ──────┐
  ──┤ Absolute Value │──
    │ Source    N7:2 │
    │              0 │
    │ Dest      N7:4 │
    │              0 │
    └────────────────────┘
```

(Parameters shown are examples
only, your data will vary.)

Figure 6-3: The ABS instruction has two parameters, source and destination

The **ACB**, or Number of Character in Buffer, instruction is used to determine the total characters in the buffer. The ACB instruction is very similar to the ABL instruction, with the only difference being that it doesn't count the end of line character. This instruction is shown below.

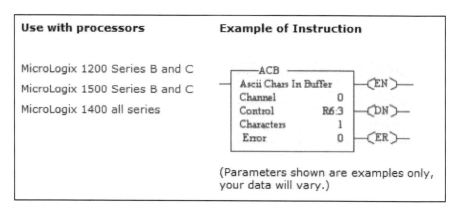

Figure 6-4: The ACB instruction

ACI (String to Integer) is used to convert a string value to an integer value that is between -32768 and 32767. To use the ACI instruction you need to have a string data type set up and a free integer data type byte. This instruction only works with numeric characters, meaning 0-9 characters. This instruction also sets the carry, overflow, zero, and sign math bits. **AIC** is the inverse instruction and converts an integer to a string. The image on the next page shows the ACI instruction.

The **ACL** (ASCII Clear Buffer) instruction is used to clear an ASCII buffer. This instruction clears the receive and/or send buffers for a certain channel. When the rung goes from false to true, this instruction will execute and clear all buffers. The second image on the next page shows the ACL instruction.

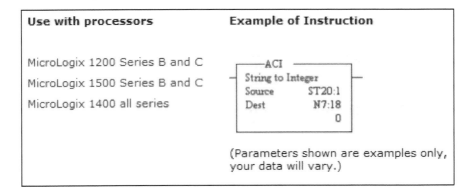

Use with processors	Example of Instruction
MicroLogix 1200 Series B and C	———ACI ——— String to Integer Source ST20:1 Dest N7:18 0
MicroLogix 1500 Series B and C	
MicroLogix 1400 all series	
	(Parameters shown are examples only, your data will vary.)

Figure 6-5: The string to integer instruction

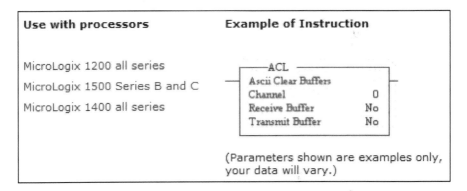

Use with processors	Example of Instruction
MicroLogix 1200 all series	———ACL ——— Ascii Clear Buffers Channel 0 Receive Buffer No Transmit Buffer No
MicroLogix 1500 Series B and C	
MicroLogix 1400 all series	
	(Parameters shown are examples only, your data will vary.)

Figure 6-6: The ACL instruction and which PLCs it can be used with

ACN (String Concatenate) combines two strings by using ASCII strings as operands. This instruction can be used with all string data types. The instruction is used to combine source A and source B into AB at a chosen destination. The image below shows the ACN instruction.

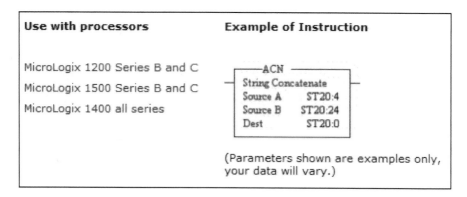

Use with processors	Example of Instruction
MicroLogix 1200 Series B and C MicroLogix 1500 Series B and C MicroLogix 1400 all series	┌─ACN ─┐ String Concatenate Source A ST20:4 Source B ST20:24 Dest ST20:0 (Parameters shown are examples only, your data will vary.)

Figure 6-7: The ACN instruction

ACS (Arc Cosine) is a math instruction used to calculate the arc cosine of the source in radians and place the result in the intended destination. Similar math instructions are the **ASN** (Arc Sine), **ATN** (Arc Tangent), **COS** (Cosine), **SIN** (Sine) and **TAN** (Tangent) instructions. All these instructions calculate the specific math function (Cosine, Sine, etc.) in radians. This instruction can only be used on MicroLogix 1400 series PLCs. The image below shows the ACS instruction.

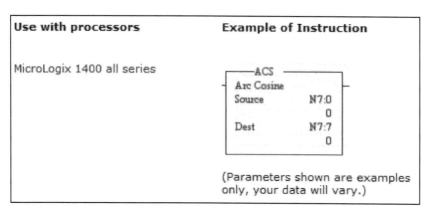

Use with processors	Example of Instruction

MicroLogix 1400 all series	┌─ACS ─┐
	Arc Cosine
	Source N7:0
	              0
	Dest N7:7
	              0
	
	(Parameters shown are examples only, your data will vary.)

Figure 6-8: The arc cosine instruction

68

The **ADD** (addition) instruction is used to add two numbers together. When the rung leading to this instruction is true, source A is added to source B and stored at the destination. This can cause some issues if either source A or source B is the same as the destination location, because it will add until an overflow error occurs. The **SUB** (subtraction) instruction operates similarly to the ADD instruction. The image below shows the ADD instruction.

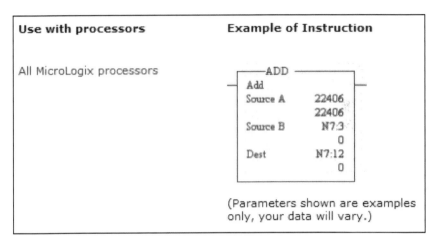

Use with processors	Example of Instruction
All MicroLogix processors	(figure)

Figure 6-9: The ADD instruction, which works with all MicroLogix processors

The **AEX** (String Extract) instruction takes a portion of an existing string and links it to a new string destination. For this instruction you provide the source string, with index being the starting point of the string you want to extract. Number is the number of characters you want to extract from the source. The image below shows the AEX instruction.

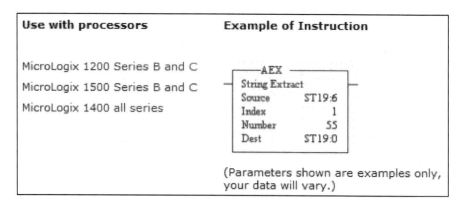

Figure 6-10: The AEX instruction

AHL (ASCII Handshake Lines) is used to set or reset the RS-232 Data Terminal Ready (DTR) and Request to Send (RTS) handshake control lines. The AND mask is used to reset the DTR and RTS control lines, while the OR mask is used to set the DTR and RTS control lines. The channel status field is used to display the status of the handshake. The image below shows the AHL instruction and which PLCs it will work for.

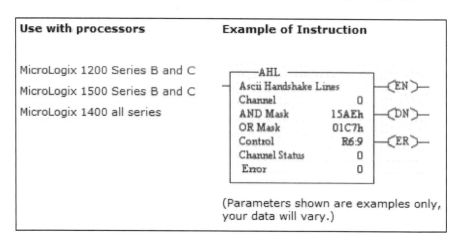

Figure 6-11: The AHL instruction

The **AND** (logical AND operation) instruction is used to combine together either a constant and a word address or two-word addresses. It then stores the result in the destination file that you choose. The **OR**, **XOR** and **NOT** instructions work in the same manner as the AND instruction as far as passing the data, but have different criteria to be true. An example is shown in the table below using four bits. The NOT instruction isn't shown in the table below, but all it does is invert the bits given from the source.

AND		
Source A	**Source B**	**Destination**
0	0	0
0	1	0
1	0	0
1	1	1
OR		
Source A	**Source B**	**Destination**
0	0	0
0	1	1
1	0	1
1	1	1
XOR		
Source A	**Source B**	**Destination**
0	0	0
0	1	1
1	0	1
1	1	0

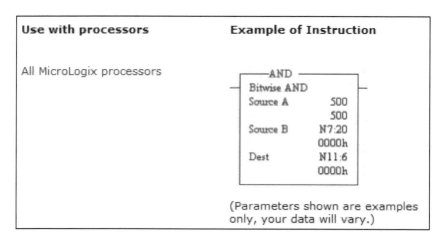

Use with processors	Example of Instruction

All MicroLogix processors

```
┌─── AND ────┐
│ Bitwise AND │
│ Source A        500 │
│                 500 │
│ Source B      N7:20 │
│               0000h │
│ Dest          N11:6 │
│               0000h │
└────────────────────┘
```

(Parameters shown are examples only, your data will vary.)

Figure 6-12: The bitwise AND instruction compares every bit from source A to source B that is in the same position and stores it in the desired destination file

The **ARD** (ASCII Read Characters) instruction is used to read characters from your chosen buffer and store them in a destination string. To use this instruction, you need to select the chosen channel to read, and set its destination. The string length field is the amount of characters that you want to read. Characters read is the number of characters that you read from the buffer and transferred to the destination. If for some reason the instruction is interrupted or canceled, the error bit will become true. This command will only execute once when the rung goes from false to true, and to execute it again the rung needs to become false again. The image below shows the ARD instruction before it has been executed.

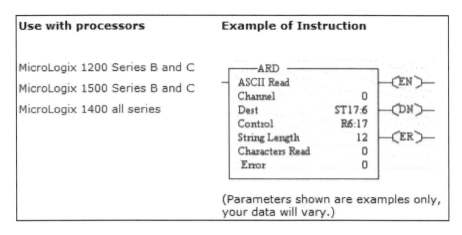

Use with processors	Example of Instruction

MicroLogix 1200 Series B and C

MicroLogix 1500 Series B and C

MicroLogix 1400 all series

```
┌────ARD─────────────┐
─┤ ASCII Read         │──(EN)──
 │ Channel        0   │
 │ Dest        ST17:6 │──(DN)──
 │ Control      R6:17 │
 │ String Length  12  │──(ER)──
 │ Characters Read 0  │
 │ Error          0   │
 └────────────────────┘
```

(Parameters shown are examples only, your data will vary.)

Figure 6-13: The ARD instruction will attempt to read 12 characters from the ASCII buffer

The **ASC** (String Search) instruction searches an existing string for a specific string occurrence such as a source string. A lot of times when a string message is sent, it has a specific indicator character(s) and a specific terminator character(s). This helps keep data separate so information doesn't get mixed up or lost. For this instruction, the source is the string you want to find, and index is the starting position when searching. String search is the string that you will search to find the source string, and result is an integer value indicating how many characters from the index the source string was found. If you wanted to search for 'P' in the string 'HAPPY' starting at index 1 (the A), the result would be 1. The image on the next page shows the ASC instruction.

The **ASR** (String Compare) instruction simply compares two ASCII strings. If source A matches source B in character length, and each position has the same upper/lower case symbols, then the rung will be true. The second image on the following page shows the ASR instruction.

73

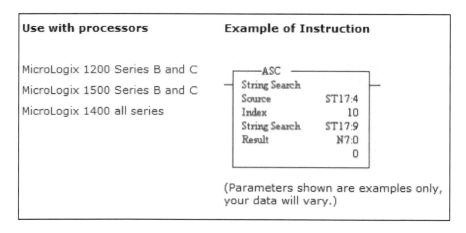

Use with processors	Example of Instruction
MicroLogix 1200 Series B and C MicroLogix 1500 Series B and C MicroLogix 1400 all series	──ASC── String Search Source ST17:4 Index 10 String Search ST17:9 Result N7:0 0 (Parameters shown are examples only, your data will vary.)

Figure 6-14: The ASC instruction is useful for making sure that you only receive the desired data

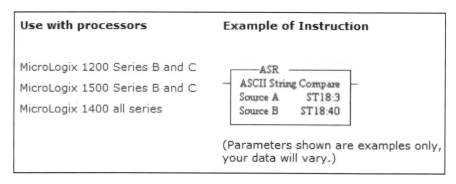

Use with processors	Example of Instruction
MicroLogix 1200 Series B and C MicroLogix 1500 Series B and C MicroLogix 1400 all series	──ASR── ASCII String Compare Source A ST18:3 Source B ST18:40 (Parameters shown are examples only, your data will vary.)

Figure 6-15: The ASR instruction is useful for making sure an input string matches your desired result

AWA (ASCII Write with Append) is used to write characters to an external device. This instruction adds the two appended characters on channel 0. The chosen channel must be channel 0. The source is the string that you want to send, and characters sent is the number of characters that the processor has sent to the device. The instruction isn't done sending data until the DN bit is set true. The **AWT** instruction works the same way as the AWA instruction, except it doesn't add two

74

appended strings. The image below shows the AWA instruction.

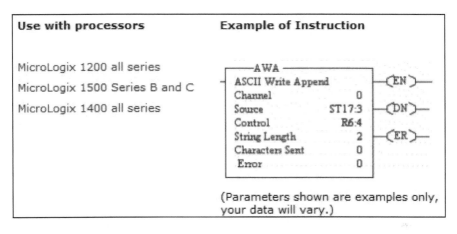

Use with processors	Example of Instruction
MicroLogix 1200 all series	
MicroLogix 1500 Series B and C	
MicroLogix 1400 all series	

AWA
ASCII Write Append — (EN)—
Channel 0
Source ST17:3 — (DN)—
Control R6:4
String Length 2 — (ER)—
Characters Sent 0
Error 0

(Parameters shown are examples only, your data will vary.)

Figure 6-16: There are several devices that require string input data and the AWA instruction is used to transmit that data

The **BSL** (Bit Shift Left) instruction is used to manipulate a bit array. This instruction takes all of the bits in a bit array and shifts the pattern through the array to the left. The first bit is replaced by a zero and the end bit is lost. The bit address field is the location of the bit which will be added to the array, and the length field is the number of bits that will be shifted. The **BSR** (Bit Shift Right) instruction is the same as BSL except it moves in the opposite direction. These instructions can be used to track parts moving through certain positions in a system. The image on the next page shows the **BSL** instruction in use.

CLR (Clear) is used to clear all of the data at a destination of your choice. This instruction is usually used with an entire 16-bit array or integer. It is also usually used at PLC startup or at the end of a certain process to eliminate all old data. This instruction is shown on the next page.

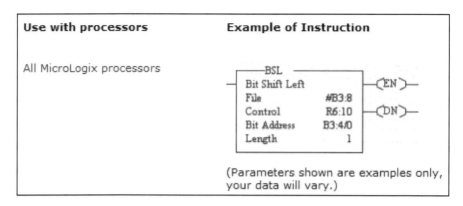

Use with processors	Example of Instruction

All MicroLogix processors

```
——BSL ——
—| Bit Shift Left
   File          #B3:8          —(EN)—
   Control       R6:10          —(DN)—
   Bit Address   B3:4/0
   Length        1
```

(Parameters shown are examples only, your data will vary.)

Figure 6-17: The BSL and BSR instructions only execute once on a false to true transition

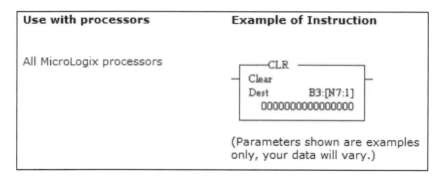

Use with processors	Example of Instruction

All MicroLogix processors

```
——CLR ——
—| Clear
   Dest       B3:[N7:1]
       0000000000000000
```

(Parameters shown are examples only, your data will vary.)

Figure 6-18: The CLR instruction

The **COP** (Copy File) instruction copies the data from the source to the destination when the rung goes true. The data is only copied for the given length or number of characters specified in the instruction command. The **CPW** (Copy Word) instruction is similar to the COP instruction, except that it is used to copy multiple bytes of data at a time. The image below shows the COP instruction.

76

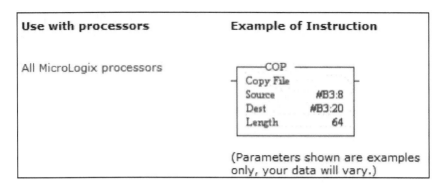

Use with processors	Example of Instruction
All MicroLogix processors	COP Copy File Source #B3:8 Dest #B3:20 Length 64 (Parameters shown are examples only, your data will vary.)

Figure 6-19: The COP instruction can be used with all MicroLogix PLCs

CPT (Compute) can be used to complete multiple math commands at a time. This instruction gives you the ability to condense multiple lines of math instructions down into one. You can use multiple data locations and do multiple different arithmetic manipulations, as long as the math expressions you enter work. The image below shows the CPT instruction.

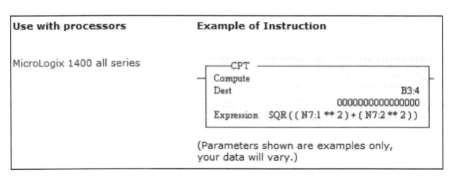

Use with processors	Example of Instruction
MicroLogix 1400 all series	CPT Compute Dest B3:4 0000000000000000 Expression SQR ((N7:1 ** 2) + (N7:2 ** 2)) (Parameters shown are examples only, your data will vary.)

Figure 6-20: Here the expression uses two data locations in its calculation

The **DCD** (Decode 4 to 1 of 16) instruction reads a 4-bit value and turns on the corresponding bit in the destination word. For example, if the four bits are 1001 then bit 9 of the destination word will be set to 1. The opposite of the DCD instruction is the **ENC** (Encode) instruction. This instruction searches the source from the lowest to highest bit for the first bit set true. That bit position is written to the destination as an integer. An example of the DCD instruction is shown below.

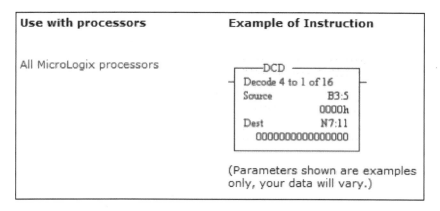

Use with processors	Example of Instruction

All MicroLogix processors	DCD
	Decode 4 to 1 of 16
	Source B3:5
	            0000h
	Dest N7:11
	     0000000000000000
	
	(Parameters shown are examples only, your data will vary.)

Figure 6-21: The DCD instruction, which can be used for rotary switches, keypads, etc.

DEG (Radians to Degrees) is used to convert from radians to degrees. After you use the COS or similar instruction the output is in radians, which is not very useable for most people. This instruction helps you convert that data to a more user-friendly form. The instruction is shown below.

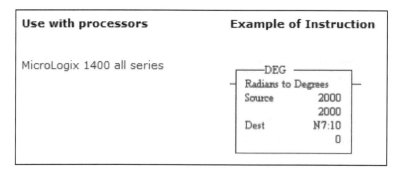

Use with processors	Example of Instruction

MicroLogix 1400 all series	DEG
	Radians to Degrees
	Source 2000
	           2000
	Dest N7:10
	         0

Figure 6-22: The DEG instruction is useful for displays and further math instructions

DIV (Divide) is another basic math function in RSLogix. When true this rung will divide source A by source B and store the result in the destination. For this instruction, either A or B or both must be a data address. The **MUL** (Multiply) instruction uses the same data types and has the same requirements as the DIV instruction. The image below shows the DIV instruction.

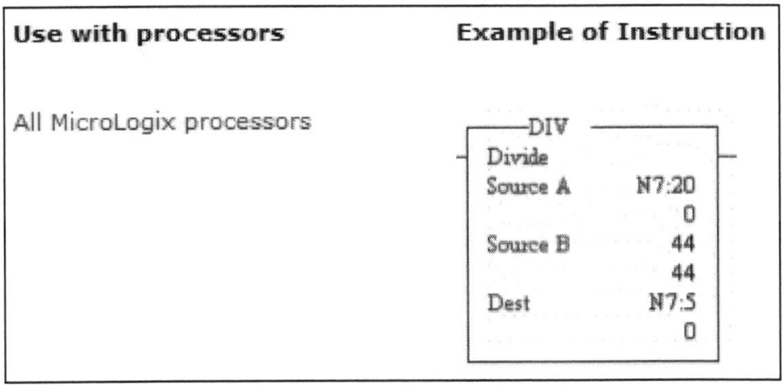

Use with processors	Example of Instruction
All MicroLogix processors	DIV Divide Source A N7:20 0 Source B 44 44 Dest N7:5 0

Figure 6-23: The DIV instruction using a data address at source A, and an INT at source B

EQU (Equal) is used to test whether two data addresses are equal or if a specific data address equals a chosen integer. If the values match, then the instruction becomes true and the next instruction can be executed. Along with the EQU bit you can use the **NEQ** (Not Equal) instruction to make sure source A does not equal source B before executing the rest of the run. Along the same lines, the **GRT** (Greater Than) and **GEQ** (Greater Than or Equal To) instructions confirm that source A is either greater than or equal to source B, or just greater than source B. The **LES** (Less Than) and **LEQ** (Less Than or Equal To) instructions are used to make sure that source A is either less than or equal to source B, or just less than source B before becoming true. The image below shows the LES instruction.

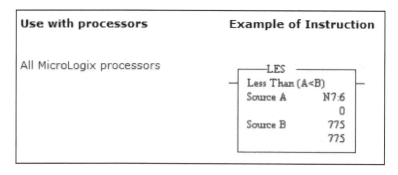

Use with processors	Example of Instruction

All MicroLogix processors	LES
	Less Than (A<B)
	Source A N7:6
	                     0
	Source B 775
	                     775

Figure 6-24: The LES instruction with source A being a data address and source B being an integer

The **FRD** (Convert from BCD to Integer) instruction converts the input source BCD into an integer and stores it in the destination. It is important when receiving input data as a BCD to convert it to an integer before manipulating it, because data can get lost if it isn't converted first. The value at the source field is the value before being converted into an INT and the value at the destination field is the value after the BCD has been converted. The image below shows the FRD instruction.

Use with processors	Special Considerations	Example of Instruction

all MicroLogix	Destination can be a word address or the math register.	FRD
		From BCD
		Source N7:5
		                  0000h
		Dest N7:16
		               0

Figure 6-25: The FRD instruction

80

The **HSC** (High Speed Counter) instruction is used to count high-speed pulses from a high-speed input. This instruction is similar to the CTU instruction and can be used for the same purpose if necessary. The HSC instruction can keep track up to 8000 Hz. The image below shows the HSC instruction.

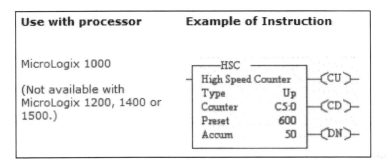

Figure 6-26: The HSC instruction for a MicroLogix 1000 PLC

The **HSE** and **HSD** (High Speed Interrupt Enable/Disable) instructions are used in pairs to help you accurately execute applications that use the HSC instruction. These instructions use the same counter data file, C5:0, as the high-speed counter. They are also the instructions that enable and disable the HSC interrupt, the counter high, the counter low, as well as the overflow and underflow states. These instructions do not work on their own without the HSC instruction being used somewhere else in your program. The image below shows the HSD and HSE instructions.

Use with processor	Example of Instruction	Use with processor	Example of Instruction
MicroLogix 1000 (Not available with MicroLogix 1200 or 1500.)	┌─HSE ─────┐ HSC Interrupt Enable Counter C5:0	MicroLogix 1000 (Not available with MicroLogix 1200, 1400 or 1500.)	┌─HSD ─────┐ HSC Interrupt Disable Counter C5:0

Figure 6-27: The HSD and HSE instructions can only be used with the MicroLogix 1000 PLC

81

Another high-speed counter is the **HSL** instruction. This is an output instruction used to set the low and high preset values for high-speed counters in MicroLogix 1200 and 1500. These PLCs have two built in high-speed counters with the instructions HSC:0 and HSC:1, with each having 36 sub-instructions available via the function files folder. A few examples would be **HSC:0/CE** (Counter Enable), **HSC:0/HPR** (High Preset Reached) and **HSC:0/HIP** (High Preset). The HSL instruction makes setting and using some of the HSC:0 and HSC:1 instructions easier to manage. The HSL instruction is shown below.

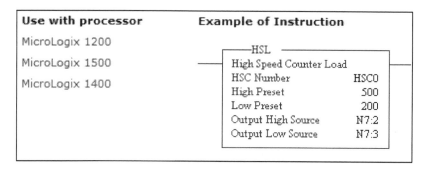

Figure 6-28: An example of how to set up the HSL instruction

The **LCD** (Liquid Cristal Display) function is for the MicroLogix 1100 and 1400 units that have built in LCD displays. Using this function, you can display strings and integer values that reference data files. There are 6 fields that you can add data to for displaying. The image below shows the LCD instruction.

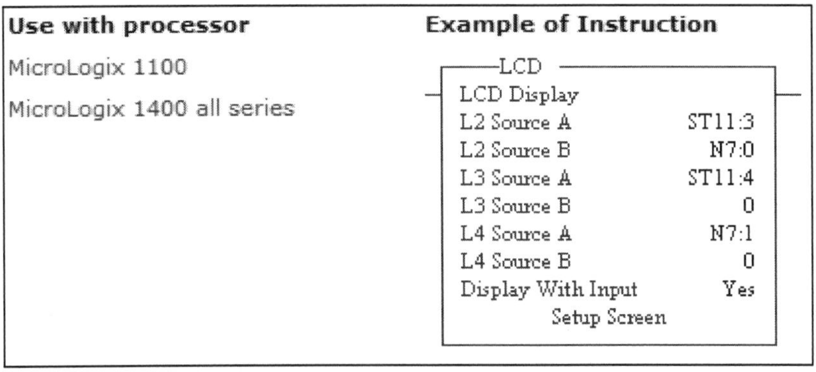

Use with processor	Example of Instruction
MicroLogix 1100	
MicroLogix 1400 all series	

```
         ─────LCD ───────
      ─┤  LCD Display                        ├─
         L2 Source A              ST11:3
         L2 Source B                N7:0
         L3 Source A              ST11:4
         L3 Source B                   0
         L4 Source A                N7:1
         L4 Source B                   0
         Display With Input         Yes
                    Setup Screen
```

Figure 6-29: The LCD instruction can use string and integer data files. You can also type in constant values to be displayed

Earlier we talked about comparison functions such as EQU, GRT and LES. There is another single function that you can use to test whether a variable is inside of a specified range. The LIM (Limit Test) takes in three parameters: low limit, high limit, and test. Each of these parameters can either be a word or a constant, but if the test parameter is a constant the other two must be a word address. The low and high limit determine the range, while the test input is the parameter that should fit inside. If it does, the LIM instruction will be true. You can also set up the instruction so that the lower limit value is greater than the higher limit value. In this case the test input has to fall outside of the range to be true. If the low limit is 12 and high limit is 10, then if test equals 11 the instruction will be false. The image on the next page shows the LIM instruction.

The **LN** (Natural Log) and **LOG** (Log Base 10) are a few more math functions that can only be used on the MicroLogix 1400 series PLC. Both functions have a source and destination parameter. When the function is active the natural log or log base 10 of the source is taken and stored in the destination folder. The second image below shows the LN instruction.

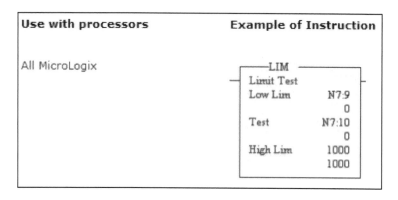

Figure 6-30: The LIM instruction with a Low LIM and Test word address, and a High Lim constant

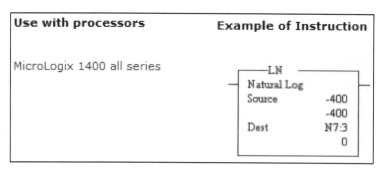

Figure 6-31: The LN instruction

You can use the **MSG** (Message) instruction to send data out to other devices through the DH-485 communication network. The MSG instruction can be used as a write or read message, and the target device can be a PLC or another device. The method depends on the device you are communicating with. This instruction is shown below.

84

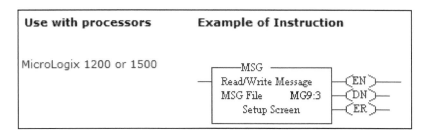

Use with processors	Example of Instruction
MicroLogix 1200 or 1500	┌─MSG ─────┐ ─┤ Read/Write Message ├─(EN)── │ MSG File MG9:3 │─(DN)── │ Setup Screen │─(ER)──

Figure 6- 32: The MSG instruction for a MicroLogix 1200 and 1500

ONS (One Shot), **OSR** (One Shot Rising) and **OSF** (One Shot Falling) are used to make sure the rest of a rung only executes either on the false to true, or true to false transition. This means that if certain instructions stayed true for longer than one scan, but the ONS instruction was used, the output would turn false after one scan. The OSR instruction functions the same way as the ONS. The **OSF** is the same concept, except it triggers the output as the rung goes false instead of true. It also doesn't depend on the rung status, but a specific input at the source and turns on a specified output bit. The ONS instruction is very useful for startup commands, to make sure the program goes to a specified state when it is first powered up and doesn't stay there afterwards. This is done by placing the ONS at the very beginning of the rung. ONS and OSR need to be assigned to unique bit addresses. Both instructions are shown below.

Use with processor	Example of Instruction	Use with processors	Example of Instruction
MicroLogix 1200 MicroLogix 1500 MicroLogix 1400	┌─OSF ─────┐ ─┤ One Shot Falling ├─ │ Storage Bit N19.0/2 │ │ Output Bit B3:2/6 │	MicroLogix 1200 MicroLogix 1500 MicroLogix 1400	B3:0 ─[ONS]─ 4

Figure 6-33: The OSF and ONS instructions

The **PID** (Proportional/Integral/Derivative) instruction is a closed loop instruction used to control aspects such as temperature, pressure, liquid level and flow rate. PID requires active analog input data, and a lot of fine tuning if you don't know the mathematics of the system. This instruction can be extremely useful for control programming, so it will be covered in more detail in a later chapter.

The **RES** (Reset) instruction is used to reset counters and timers. This is an output instruction and has to be used at the end of a rung. A counter or timer does not have to be finished in order to reset them. The image below shows the reset instruction.

Processors	Example of Instruction
All MicroLogix processors	C5:9 —(RES)—

Figure 6-34: The reset instruction

The **SCL** (Scale Data) and **SCP** (Scale with Parameters) functions are used for scaling analog input data into something more useful. Most analog input data is either 4-20mA or 0-10V, which transitions to a value between 0 and 32767. This value is not very useful for operators or even programmers to read. Using the SCP instruction you can set the input, the input min and max, as well as your desired scaled min and max. The instruction then scales the data and sends it to your desired output word address. This function is very useful to use with the PID instruction to make the data easier to analyze from the operator's point of view.

86

The SCP instruction is slightly easier to use than the SCL instruction, which is why it's covered instead of SCL.

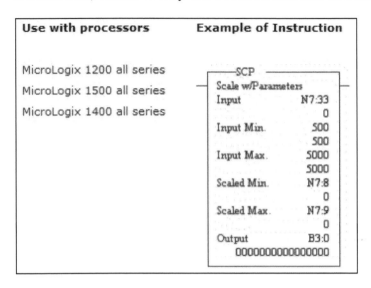

Figure 6-35: The SCP instruction

Another set of math instructions that can be useful for programming, is the **SQR** (Square Root) and **XPY** (X to the Power of Y) instructions. There are a couple of important things to remember when using these math instructions. If you are using the SQR function and the value cannot be squared to a hole number, the PLC will round the final result. If you are using the X to the power of Y function, remember that it is very easy to create numbers that are too large which can cause an overflow. This will result in data that isn't very useful. The XPY function takes source A to the power of source B and stores it in the destination, while the SQR function simply takes the input and creates the square root.

SUS (Suspend) is a useful instruction for debugging your program while it is running. When the suspend instruction is true, the controller goes into an idle state and all outputs are de-energized. However, you can still see the state of every word address and bit address in the PLC. You could also set up your program so that if a certain condition occurs, the PLC goes into a suspended state to give you the chance to see what went wrong. This instruction is shown below.

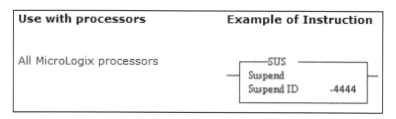

Use with processors	Example of Instruction
All MicroLogix processors	SUS Suspend Suspend ID -4444

Figure 6-36: Each SUS instruction can be given an ID to help you know which is active

When dealing with data transferring, you have to be conscious of the order that the data is sent in. Depending on the device and the method for data transfer, certain words of data can be swapped backward from what is needed. If you are reading two numbers off of a device, and you know that the two numbers should give you an output of 20,18 but you receive 18,20, it could be because the data you received needs to be swapped. To do this you can use the **SWP** (Swap) instruction, which swaps the first half of data with the second half of data. This instruction is shown below.

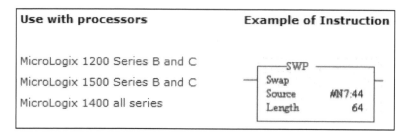

Use with processors	Example of Instruction
MicroLogix 1200 Series B and C MicroLogix 1500 Series B and C MicroLogix 1400 all series	SWP Swap Source #N7:44 Length 64

Figure 6-37: To use the swap instruction you have to give a word address, and the length of data you need swapped.

88

This brings us to the end of the chapter, and the **END** instruction. The END instruction is automatically placed at the end of every program file and cannot be edited. There are many useful instructions that you can use when creating your program, and many of these instructions overlap in their capabilities. The instructions covered in this chapter will get you through most of your programming problems. Many of the most useful instructions have specific sections dedicated to them later in this guide, which should clear up any questions you might have at this point.

7. Timers, Counters and Integers

In this chapter we will discuss how to use the timer and counter instructions, as well as look at examples of each. We will also go more in-depth into how integers work and look at how data that is too large for a single word address can be split to multiple word addresses.

7.1 Timers

In RSLogix there are multiple timer instructions that you can choose from, and each of them is an output instruction. There are three different types of timer instructions: the TON, TOF, and RTO. The **TON**, or Time On Delay, instruction begins timing when the rung goes from false to true, and only tracks the time that passes as long as the rung remains true.

TOF is the Time Off Delay instruction. For this timer to start counting, it has to go from a true state to a false state. Like the time on delay instruction, the TOF instruction will start over if the state goes back to true. The **RTO** (Retentive Time On Delay) instruction is similar to the TON instruction because it acts like an on-delay timer, but unlike the TON instruction if the rung goes false the accumulated time is retained until you reset it, or power is cycled. The TON instruction is shown below.

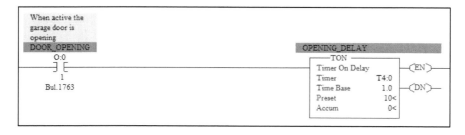

Figure 7-1: Timer on delay that will run for 10 seconds before it is complete

Timers in RSLogix also give you the ability to select the time base as either a 1.0 second, .01 second, or .001 second time base. So depending on the application and how long something will need to run, you can choose the most effective time base. Usually you would choose the millisecond time base to get the most accuracy, but this is not required. The number you enter into the preset position is multiplied by the time base to give you the total number of seconds the timer will run until it is done.

Each timer has an EN (enable bit), TT (timer timing bit) and a DN (done bit) to use at with the chosen data type. If you are using the T4:0 data type, T4:0/EN is true if the timer instruction is running. You can use the T4:0/EN instruction if you want to confirm that the timer is on, but typically this is not needed. T4:0/TT is only true when the timer is counting. Once the timer hits the preset time, the TT bit becomes false. This is useful for giving certain lines of code only a specified amount of time to run, or to set up your code so that once the timer begins it will run the full timer. The T4:0/DN bit only becomes true once the timer has finished counting and is still enabled. An example of T4:0/TT and T4:0/DN is shown below.

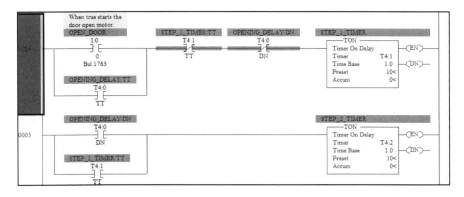

Figure 7-2: Multiple timers to separate different states

Now that you know how set up a timer, we will cover some of the more popular uses for timers. One of the most used practices is to account for the debounce time. A debounce timer is used to keep an output from flickering on and off numerous times due to real-world input. If you are using a PLC to control the level in a tank, you will want to use a debounce timer. This is because when you are draining or filling up a tank, the water will slosh around causing the water level sensors to turn on and off many times. This will continue until the water level is high or low enough. We need the input to remain on or off for a few seconds straight, so this instability in the levels do not affect the output. This will also reduce the wear on the physical parts that fill or drain the tank due to rapid stopping and starting.

Keeping with the tank fill example, you can also use timers to activate alarms. You can set up a timer to start when the tank level hits the low mark, if the goal is to fill the tank back up. The program will then either open a valve or turn on a pump. If the tank level doesn't rise high enough to turn off the low-level sensor after several seconds, then you know the tank is not filling as intended. When the timer finishes you can have the DN bit trip an alarm to indicate that something is wrong.

Another great use for timers is to space out the execution of different steps in the code. If you need to ensure that two moving parts don't collide, you can use a timer that starts once part one starts moving. Part two then has to wait a specified time before it starts moving. In another scenario, if you know it takes a component five seconds to travel to the end of a conveyor system from a certain position before it is picked up, you can use a timer to confirm that it arrives as planned. If the component doesn't arrive on schedule, the pickup unit could malfunction.

For our garage door example, there are several places where we could use timers to help the program run more smoothly. The first instance is a timer to ensure the door opens and closes in the correct amount of time. After we instruct the garage door to start opening, there is no way of knowing if the door is actually moving in between the door closed and door open sensors. Instead of assuming that the motor is still working, the chain didn't break, or the door didn't get stuck, we will need to measure how long the door takes to open (add a couple seconds to allow for error in measurement) and make a timer that starts timing once an open/close request is made. An example of this is shown on the next page.

Imagine that our door opened properly, and a vehicle is driving into the garage. The vehicle detection sensor might only be high enough to see the tires, so once the front tire passes the sensor the garage door closes on the car as it is driving. Since the sensor saw a car enter and leave, the program will think that the entire car is clear.

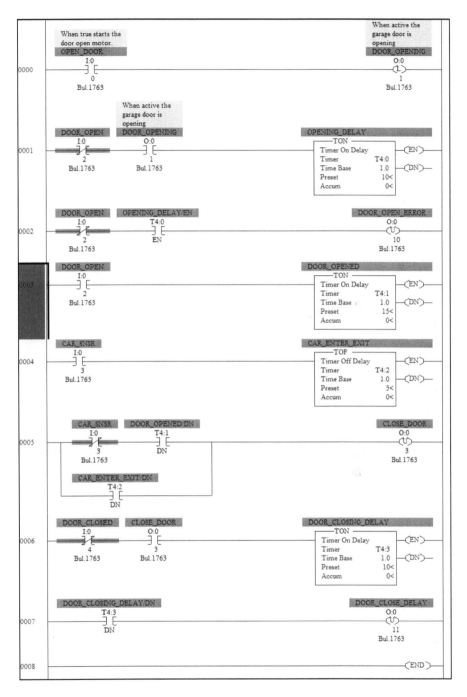

Figure 7-3: An error timer on the open and close commands, along with a TOF timer for after the car enters the garage.

To account for this, we want to give the drivers an ample amount of time to make sure they can get through the door without having to worry about it closing on them.

In another scenario, what if someone hits the open button on accident and doesn't notice that the door is already open? We don't want the garage door to remain open indefinitely, so we want another timer that starts timing once the door is open. If this timer finishes, we should be able to assume that no cars are coming in and close the door.

Situations where you might want to use a TOF timer, would be for avoiding collisions or water hammers in pipes. If you have a pipe system in a plant being used to pump water and suddenly shut off the water supply, it can cause hydraulic shock from the water's momentum. To avoid damaging valves, you can use a TOF timer that starts timing once the pipe turns off. After the timer finishes use the TOF/DN bit to close the valve (of course this is situation based depending on how many valves there are and how the line is set up).

RTO instruction are very useful as well. Let's say you have a paint application where a component has to spend 5 minutes in the paint solution, but must be pulled out every 30 seconds in order to be turned and put back in. If you set up a normal TON timer to track how long the component is in the paint, it will restart every 30 seconds. Using an RTO every time the component is dipped back into the paint, the timer will pick up where it left off until you hit the desired 5-minute mark.

7.2 Counters

In an earlier chapter we learned that the **CTU** (count up) and **CTD** (count down) instructions count on a false to true transition. Unlike timers they only count on the transition, they don't keep counting if the rung remains on. The count up and count down instructions are essentially the same instruction, with the only difference being whether or not you want to count up from zero or down to zero.

The bits that go along with the up and down counters are the EN (enable) and DN (done) bits. The counters also have a preset value and accumulator value. The preset value is the desired number of counts before the counter is done. And the accumulator value is increased by one with each false to true transition. Unlike the timer instructions, the counters will not reset back to zero by themselves. You have to use a reset instruction and match the instruction word address to the counter you wish to reset. An example of the counter instruction with a reset instruction is shown below.

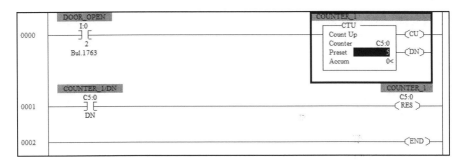

Figure 7-4: The count up instruction with the done bit which resets the counter

Counter instructions are very useful for checking for errors, batching and keeping track of positions. For tracking errors, you can count how many times certain parts of your code are executed. This is useful for making sure that the program doesn't get stuck in a loop trying to execute the same line of code over and over again. This can happen if you are using a proximity sensor and for some reason either the sensor fails, or the machine is blocked from moving to the desired location. Likewise, if you are controlling a machine that rotates and you need to count the rotations, you can make sure that it doesn't rotate too many times.

Counters are also perfect for packaging lines. If you are programing a packaging line of shampoo bottles, and you are focusing on the batching process, every time a bottle passes the end of a conveyor the accumulator can increase by one until you hit your desired batch size. Once a batch is complete, reset the counter and start over.

A less common use for counters would be to track the position of something like a turn table. If you have a turn table set up to rotate in quarter turns, the counter can increment by one every time the turn table rotates. You can then use comparison instructions to check the accumulator value versus the desired indicator value. Once the counter is done, you know a full rotation has been completed.

Using our garage door example, a counter can be used as an error counter to make sure that the door isn't opening and closing needlessly. If the garage door opens and closes 10 times, then an alarm will go off to indicate that the garage door car sensor might be damaged and should be checked. An example of this is shown below.

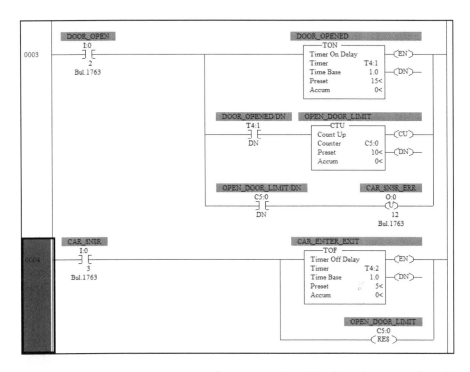

Figure 7-5: An up counter counting every time the door is opened, and resets if the car sensor turns true

7.3 Integers

The ability to manipulate numbers in RSLogix is one of its most powerful uses. This chapter is only going to cover how integers are stored in word address and how they can be broken down into individual bits. Math instructions will be covered in the next chapter.

In Chapter 5 we learned that a word is 16 bits long and can have a value between -32,768 and 32,767. The number range originates from the binary number system. A word equals bit 0 to 15, because everything starts at a zero base in RSLogix. The 15th bit is the negative bit for a normal word.

This means that if bit 15 is set equal to 1 the entire word value will be negative, and if only the 15th bit is set to 1 the word value is -32768. The table below shows a few different word values.

Binary value	Decimal value
0111 1111	32767
1000 0000	-32768
1111 1111	-1
0000 0001 1000 0000	32768

If you are using numbers larger than 32767, the data will be stored into the next word up. If your desired data word is N7:0 and the value you send to that location is too large, then some of the data will be stored in N7:1. This can cause trouble if you are not paying attention to where you are sending and storing data. This is because you can overwrite data accidentally by trying to store a value at N7:0. But if the value is to large, no matter what data you have in N7:1, it will be overwritten.

It is important to note that when creating an integer datafile, you can choose the number of elements in that data file. If the elements selected is equal to 1, then you only have access to NX:0. The more elements, the more data the file will need. This is shown below.

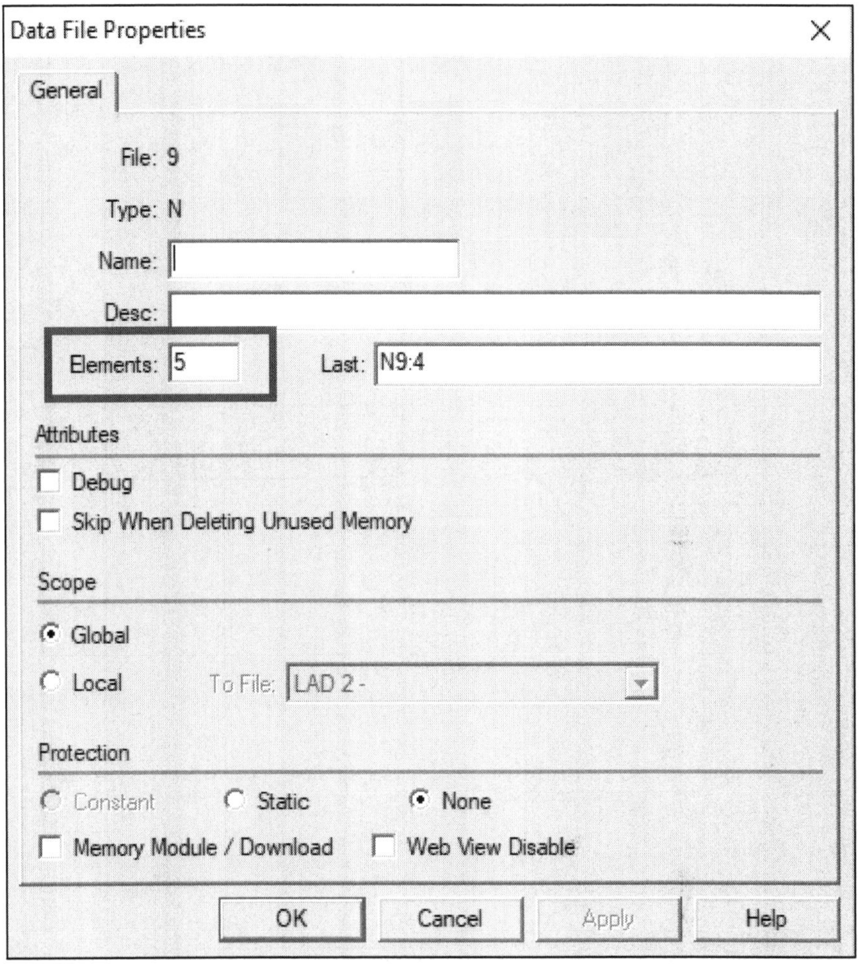

Figure 7-6: Data file properties with the elements selection highlighted

We covered how to use timers and counters in different situations, and how the different bits of each of these instructions work. This chapter also covered negative values with integers, and how larger numbers are stored into multiple word locations. The next chapter will go into more detail on math instructions and how to manipulate integers.

Learned something new?

If you found any part of this guide helpful so far, or learned something new that you can't wait to try out, I would love to hear about it. The main way for me to connect with you, is through a review on Amazon. So head over there and let me know which topics you liked most, or even which ones you didn't like. Are you using this guide in your studies or job? I would love to hear about it as well.

8. Move, Jump and Math Functions

This chapter covers instructions for moving integer values into different word addresses, and instructions to compare a word address to a default value or another word address. You will learn a few different uses for each of these instructions, including how to set up your program to guarantee the process happens in the correct order. You will also learn about jump and subroutine instructions, which allows you to jump from one section of the program to another without allowing the PLC to scan through every rung. Finally, we will cover simple math such as addition, subtraction, multiplication, division, the compute function, and finally PID control.

8.1 Move and Compare Instructions

Chapter 6 covered what each of the compare instructions are. This chapter will go further and discuss more about the uses of compare functions. If you are writing a program to control any type of process control (such as tank filling, water flow, temp control, or batch size) you will most likely need to compare values to make sure they are correct for the process. When you are monitoring the level of a tank, and if you need to know when to turn on the pump to start filling that tank, you will have to compare that tank level to some default value in order to know if the tank level is too low, too high or still in the safe zone.

Compare instructions are useful for comparing feedback to desired values for many different situations. The tank example above works as a general example for any analog input value you will be using for your project, but it is also useful for comparing states and positions of gantries and robots. A gantry is essentially a crane that is supported on four corners and can move to any location inside the four corners. For our examples, a robot means a singular robot arm that can move in every direction to pick up items.

These devices can typically be set up with a certain number of set positions. For instance, the gantry can only go to ten different locations. The only way for your code to distinguish which location the gantry is in, is to use compare functions. The image below shows an **EQU** function used to test the gantry position. For all compare functions, source A has to be a word value and source B can be either a word value or constant value.

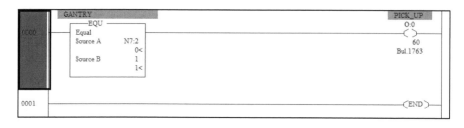

Figure 8-1: Using word address N7:2 for the gantry current position and testing to see if that value equals 1

Compare instructions are also useful for alarm and error checking. Using the **LIM** instruction is a good way to ensure everything is in the correct state for a certain section of your code to be able to run. Using the gantry example, you would always expect the values to be between 1 and 10, but what if something happens to corrupt the data? Your program would have no way of knowing that something is wrong, and any number of problems could occur. To avoid this, you would use the LIM instruction, which is the same as using the "less than or equal" and "greater than or equal" instructions in series. This instruction will ensure the Gantry position is in one of the ten designated positions it should be in, and not somewhere outside the boundaries. An example of this is shown below.

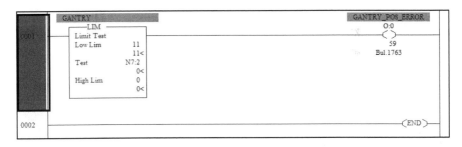

Figure 8-2: Since the Low Lim value is greater than the High Lim value, this instruction will go true if the gantry value is outside of 1-10

In order to compare values, you will have to add values into word addresses. There are multiple ways to load values into different word locations. All math instructions have a destination parameter that loads the result into your desired location, but math instructions only deal with integer and float values. Another way to load values into a specific location is to use the **MOV** instruction. When this instruction is executed, it moves either an integer, float or string value into a word address.

The data that you want to transfer has to match the data type of the word address, otherwise RSLogix will give you an error when you try to verify the program. The move instruction also gives you the ability to copy a value from one location to another (there is also a copy function as well). If you need to manipulate raw data from an analog input, you first need to move the raw data into a word address.

One of the best uses for the MOV and EQU instructions is to control your program step by step to make sure nothing happens out of order. This isn't useful in every situation, but it is for controlling specific processes of code. An example of this would be if you want to control a gantry and you know it has to go from position 1 to position 3, stopping at each position along the way. You could control it step by step as follows:

1. At startup move the value 0 into word address N7:1 (Gantry_Control_State)

2. If Gantry_Control_State = 0 and the gantry is in POS 1, then lift the gantry arm up and request move to POS 2. Move the value 10 into Gantry_Control_State

3. If Gantry_Control_State = 10 and the gantry is in POS 2, then extend the gantry arm down. Move 20 into Gantry_Control_State

4. If Gantry_Control_State = 20 and the gantry is down, then close the gripper. Move 30 into Gantry_Control_State

5. If Gantry_Control_State = 30 and the gripper is closed, raise the gantry arm up and request move to POS 3. Move 40 into Gantry_Control_State

6. If Gantry_Control_State = 40 and the gantry is in POS 3, extend the gantry arm down and open the gripper. Move 50 into Gantry_Control_State

7. If Gantry_Control_State = 50 and the gantry arm is down, raise the gripper and request move to POS 1. Move 0 into Gantry_Control_State

The sections above show a very simple example of controlling a process. Of course you will still need a lot of code outside of the process control, to make sure all the data is collected and sent out properly and there are no other errors occurring. The image below shows the above process in ladder logic.

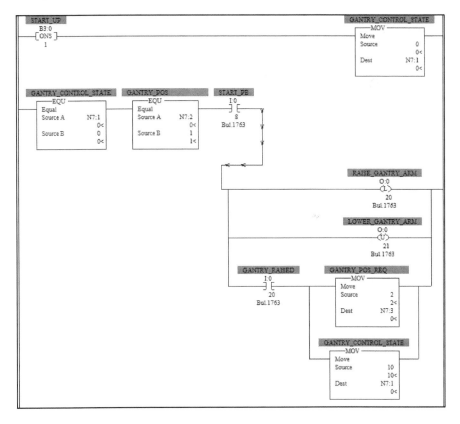

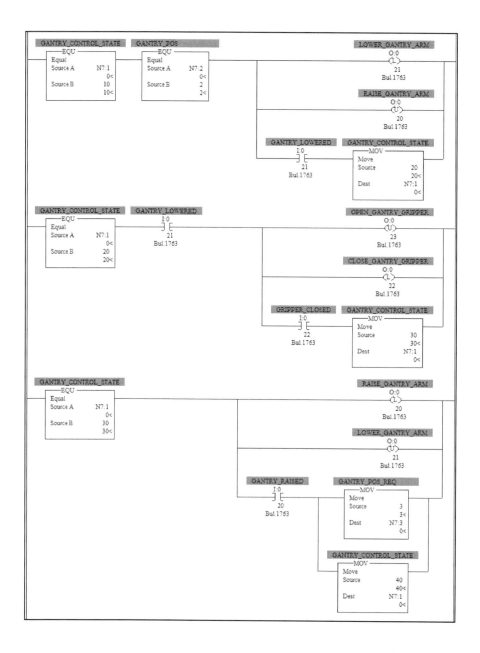

108

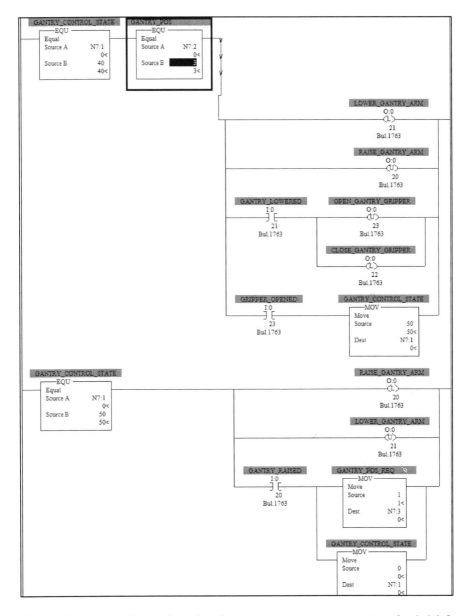

Figure 8-3: Steps through a simple gantry move process. For the initial setup, allow space in the beginning in case you need to add in steps later

8.2 Jumps and Subroutines

A **JMP** (jump) instruction is used to jump forward over segments of code, or to jump backward in the program to allow you to loop through segments of your code. A jump code is an output instruction that executes when the rung is true. Each jump instruction needs to be matched to an **LBL** (label) instruction. The label instruction is where the jump instruction will end up, and multiple jump instructions can go to the same label. Make sure to place LBL instructions at the beginning of the rung they are used on.

Jump instructions are very useful for situations where you might not know for sure which part is going to arrive on a conveyor. For example, if you are bottling different kinds of soda of different sizes, you could use jump codes to only execute small segments of your code. If bottle 1 is supposed to be soda A and it is a large bottle, your program can jump to the large soda filling segment of code and skip over the small soda filling segment. The same goes for pouring soda A and skipping the Soda B code. You should still include all of the conditional tests you normally would to confirm the bottle is indeed a large bottle etc. The image below shows a small example of using a JMP and LBL instruction.

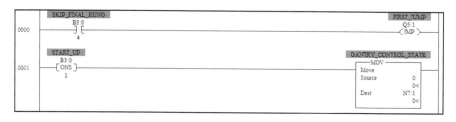

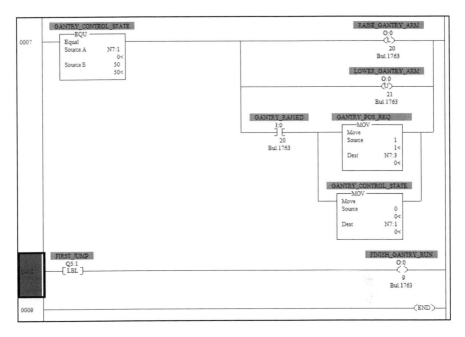

Figure 8-4: A jump instruction from rung zero down to rung eight. None of the program between the jump and label instructions will be executed

Jumps are good for doing small loops of code or jumping to specific spots in a program. Using subroutines are useful for saving memory and for saving scan time. Similar to the JMP instruction, the **JSR** (jump subroutine) instruction jumps from one spot in the program to another. The difference between the jump instruction and the jump subroutine instruction is that the JSR jumps to a completely new program designated by an SBR. The **SBR** (subroutine) instruction has to be placed at the beginning of a program file, with the rest of the program being executed afterward.

In RSLogix, only the main ladder program (LAD 2-Main Program) will execute in the normal cycle. Every other ladder you make will need to be jumped to using JSR and SBR

111

instructions. At the end of each program, the scan jumps out of the subroutine and back to the main program. Typically, the main program is used to jump to each sub program as shown below.

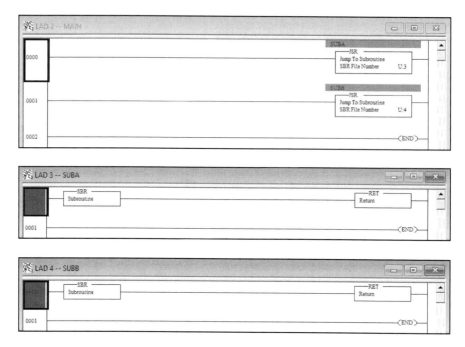

Figure 8-5: Shows the main program with two JSR instructions and two subroutine programs

8.3 Simple Math Instructions

There are many different math instructions to use in RSLogix. This section is only going to go over addition, subtraction, multiplication and division. The goal with this section is to give you a better sense for how RSLogix handles the data. As previously discussed, MicroLogix PLCs default to 16-bit math, meaning any number bigger than 32,767 or less than -32,768 can't be used. If you add two numbers together to get a larger value than 32,767 you will get an overflow error. Likewise, if you subtract two numbers and get a value less than -32,768 you will get an underflow error. If you are using a MicroLogix PLC you have the ability to do 32-bit addition and subtraction, by sending the math instruction result to S13. The resulting value can be between -2,147,483,648 and 2,147,483,647.

When using math instructions, and you know the value you want has an upper limit of 1,000, it is still important to check the overflow bit S:2/14. If you are adding two variables together and the destination is N7:1, and the result is 60,000, S:2/14 will go true. The value that N7:1 will show is 32,767. This happens because the data gets truncated and the true result will be lost. This is where scaling comes in handy to make sure data doesn't get lost during math operations.

If you are using an add or multiplication instruction where source A is a data location such as N7:0, and the result is also being stored in N7:0, the result can quickly get into the hundreds of thousands because of how fast the PLC scan time is. If you have an analog input, the direct data your PLC will receive is typically between 0 and 32767. Depending on the device you have, you will usually want to scale it between 0% and 100% for elements such as tank levels or to match the upper and lower limits of a temperature probe. Most values that get into the tens of thousands aren't very user friendly and should be scaled anyways.

The goal of this chapter was to help you learn methods for controlling your program. Move and compare instructions are useful for making sure you have the correct data where you need it, but they can also be used to make sure certain processes that need to occur step by step, happen in the correct order. Jumps and subroutines are very useful for saving time on your program and saving memory, so your program doesn't become too large to compile.

The important thing to remember about math instructions is to pay attention to the data you are manipulating.

9. Peripheral Devices

This chapter will cover communication with additional devices through ethernet IP. We will first go over changing your computer's IP address, and how to set the desired IP address in RSLogix 500. Then we will cover using the RSLinx software and how to configure new drivers. Finally, we will cover a little bit of FactoryTalk View Studio and some of the things you need to consider when setting up a communication network with an HMI.

9.1 Matching IP Addresses

The IP in ethernet IP stands for Internet Protocol. This communication method is typically used via the ethernet port on the devices you are using. But if your devices have wireless capabilities, it can also be done via the internet. An IP address is a code that follows the structure of XXX.XXX.XXX.XXX with each set of XXX equaling a number between 0 and 255. Most devices have a default IP address that is something similar to 192.168.XXX.XXX. All cell phones and computers have IP addresses.

If the PLC you are using has an ethernet port to establish communication with your device, you will first have to look up the default IP address from the manufacturer. Some devices with LCD displays or with rotary dials can display the default IP address.

You can even set the IP address through the device in some cases. However, in most situations you will need to change your laptop or computer's default IP address in order to communicate with your intended device. To do this follow the steps below.

1. In the Control Panel click on the Network and Internet tab.

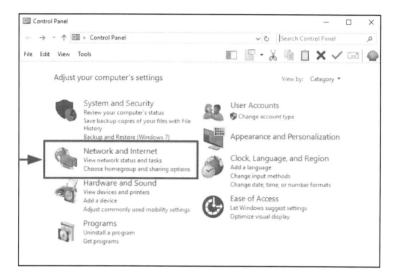

2. Then you need to navigate to the Network and Sharing center.

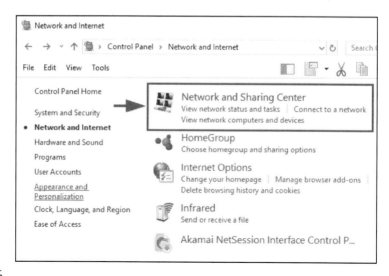

3. Once in the Sharing Center click on the Change Adapter Settings tab.

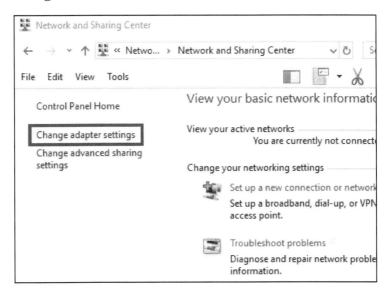

4. From here you will need to navigate to your ethernet properties menu by double clicking on the Ethernet tab.

5. In this menu you will need to click on the Internet Protocol Version 4 (TCP/IPv4) tab and click on Properties.

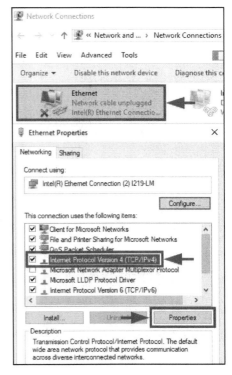

6. In this menu you can set a static IP address for your computer.

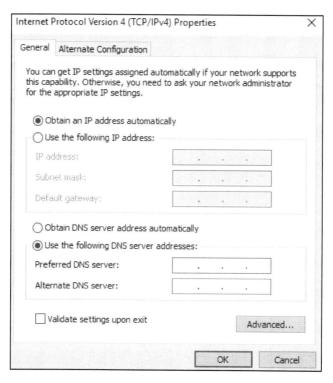

7. In order to set the IP address for your computer, you have to click the OK button and close out of the ethernet properties menu.

When setting the IP address for your computer, the first three sets of numbers need to match the device or devices you want to connect with. If your PLC has and IP address of 192.168.20.1 then your laptop will have to be set to something like 192.168.20.100. The first three sets of numbers need to match, causing the devices to be on the same network. The final number needs to be different otherwise you will cause a communication conflict. The same goes for something like an HMI on the same network.

In RSLogix 500 click on the Channel Configuration tab to bring up the channel communication settings. This screen will differ depending on which PLC you have and on the available communication methods. In the image below, channel 1 is set up for ethernet. Make sure the BOOTP Enable is unchecked, then you can set the IP Address and Subnet Mask.

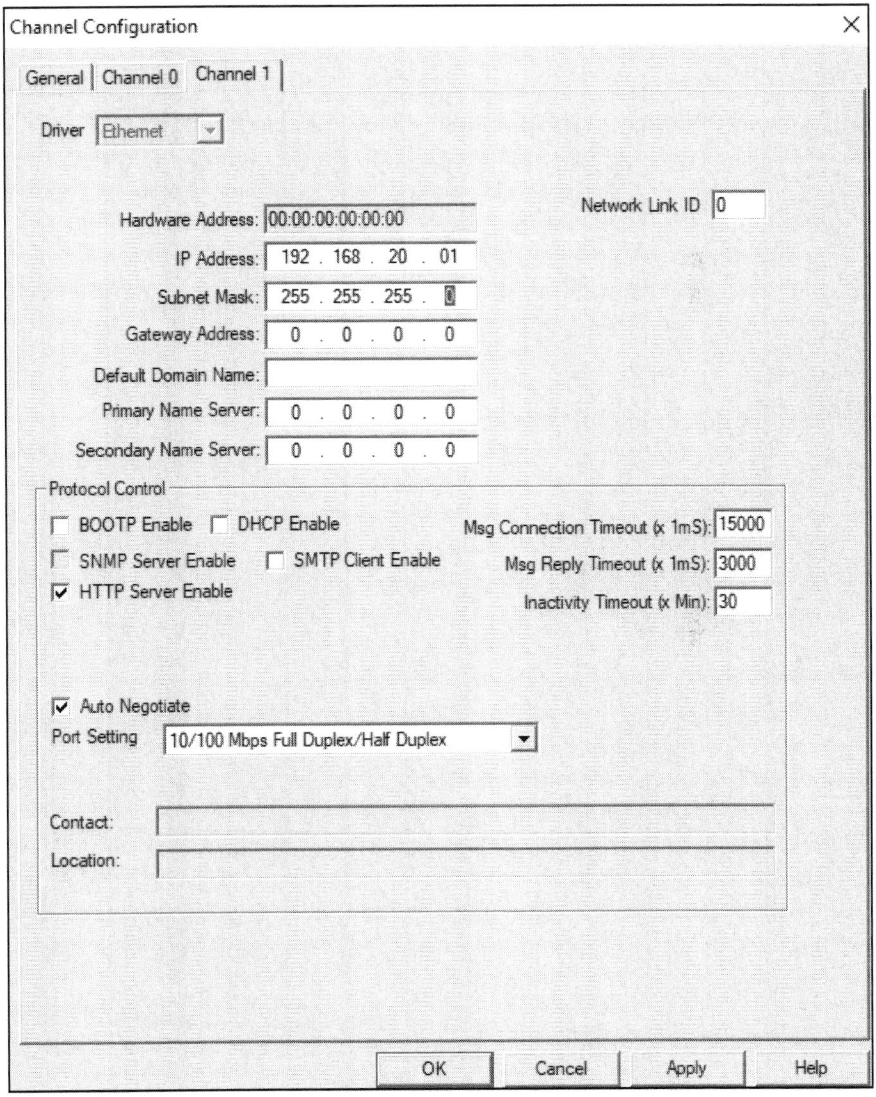

Figure 9-1: IP address set for the project as 192.168.20.1

9.2 RSLinx Classic

When you download RSLogix 500 you will also need to download RSLinx. This software is used to create connections between your computer and Rockwell devices. RSLogix 500 won't be able to see a PLC without RSLinx, no matter what the IP address is. The same goes for FactoryTalk software when trying to connect to an HMI.

RSLinx Classic Lite is available with certain Rockwell software such as RSLogix 500, Studio 5000 or FactoryTalk. RSLinx enables you to connect with multiple different devices through several different methods. This guide will only cover IP address connections via ethernet with MicroLogix devices. There are versions of RSLinx that can be bought separately, but unless you need to set up an OPC, it is not likely you will need to purchase this software. The image below shows the RSLinx Classic Lite window.

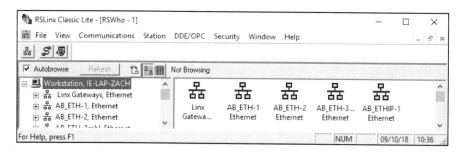

Figure 9-2: Opening screen for RSLinx

In the image above there are several driver gateways set up. These are the portals that enable your computer to detect devices such as PLCs and HMIs. The following steps will go over generating a new gateway.

1. Hover over the Communications tab in the navigation bar and click on the Configure Drivers tab.

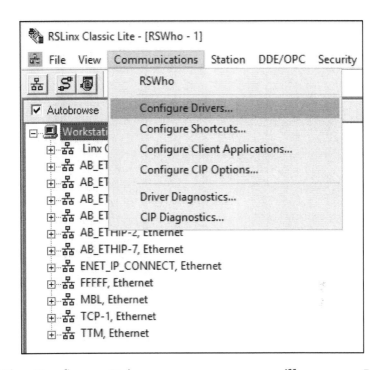

2. The Configure Drivers pop-up menu will appear. In the image below there are already several drivers configured, but this will be different depending on the drivers you have created. To configure a new driver, click on the drop-down arrow in the Available Driver Types box.

3. In the drop-down menu you need to select the EtherNet/IP Driver then click on the Add New button.

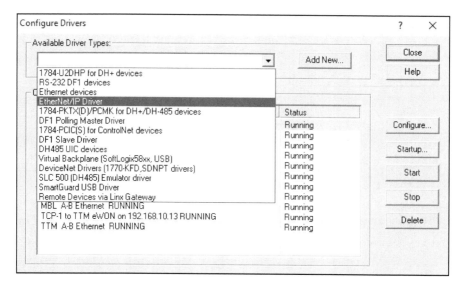

4. The Add New RSLinx Classic Driver pop-up will appear. In this menu you have to assign a unique name to the specific driver you are setting up. Then click OK.

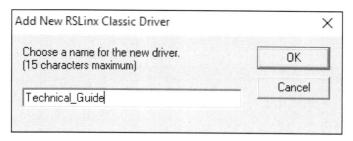

5. Next, the Configure Driver pop-up will appear. You will have two options, either to Browse Local Subnet or Browse Remote Subnet. When you are trying to connect your computer to a standalone PLC network, you need to select Browse Remote Subnet. From there you must type in the IP address of the device you are trying to connect to. For this instance we will say the PLC has the IP address of 192.168.20.1 and has a subnet mask of 255.255.255.0.

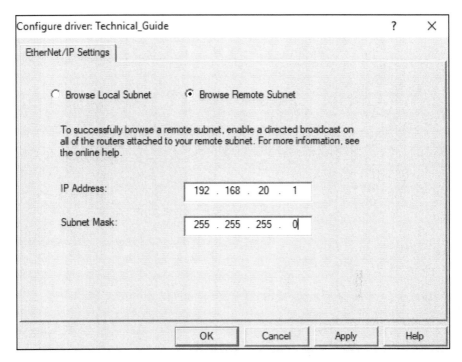

6. Now in your Configure Drivers menu you should be able to see the new driver you just created and what type of driver it is.

After you follow these steps, and your computer is set to an IP address that matches the driver you just configured, you should see your device on the main page. If your computer can no longer see a device after you have connected to it, a red X will appear over the device. In the third step above, you could see that there are many more methods of scanning for devices than just ethernet IP. Each of these drivers are set up in a different way, and have different applications depending on the device you use. The image below shows you how a device will appear in RSLinx if your computer can actively see it.

123

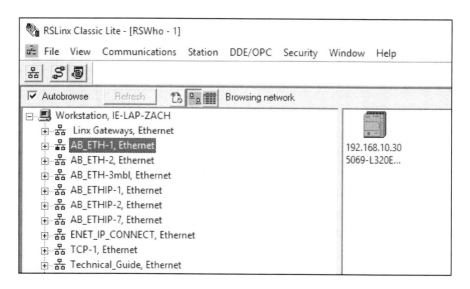

Figure 9-3: RSLinx Classic main page with a PLC connection active

9.3 FactoryTalk View Studio

FactoryTalk View Studio is an HMI building software that is often used in tandem with RSLogix 500 and 5000. The reason for this is because using these two softwares together on Allen Bradely products, allows for easier communication across different platforms. That being said, you do not have to use an Allen Bradley PanelView HMI with a MicroLogix PLC, but that will not be covered here. This chapter isn't instended to go over how to create HMI screens or how to set the IP address and tags in FacotryTalk studio. This section is intended to emphasise that for your PLC to communicate with multiple devices, all devices need to be on the same IP address network. The FactoryTalk opening window is shown below.

FactoryTalk View Studio has several default programs that is starts with. This guide will use the InstantFizz_ME application.

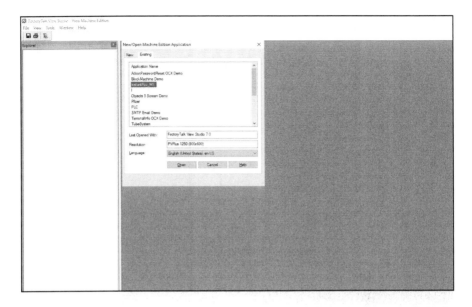

Figure 9-4: The page that opens when you open FactoryTalk View Studio

After the application opens, an explorer menu will pop up on the side bar. Under the Display section you can create the HMI screens that will appear on the PanelView device. The image on the next page shows the explorer bar with a display window opened.

The reason why we are discussing FacotryTalk, is to go over how to set up tags for communicating with RSLogix 500. In the Explorer menu there is a section called HMI Tags, which is used to control each part of the HMI screen. These tags can be matched to the tags you want to use in your PLC.

For example, if you create a start button in your HMI, you could create a tag called Start_PB and link it to one of the internal bits in your PLC. B3:0.0 could be linked to the start button tag, which would then enable you to initiate a program on the PLC through an HMI.

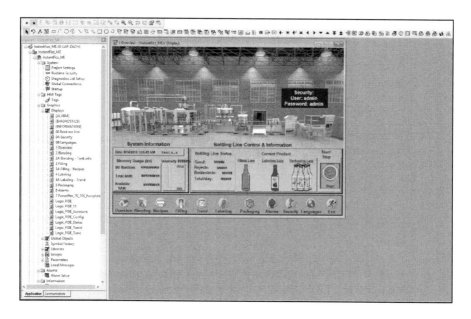

Figure 9-5: An HMI screen created using pre-defined symbols and custom drawings, with text boxes and tag connections

Matching HMI tags to bit addresses in your RSLogix 500 project isn't enough to transfer information back and forth from the HMI to the PLC. You also need to make sure that the the HMI is on the same IP network as your PLC. Currently our PLC is set to 192.168.20.1. So we have to set the HMI to 192.168.20.2 through 192.168.20.255, and connect the two devices via an ethernet cable.

There is a Communication tab in FactoryTalk studio that allows you to set up which IP address the project is going to be downloaded to. You can set up multiple HMI devices to communicate with one PLC, as long the tag addressing is set up correctly.

To recap, this chapter covered how to configure communication via ethernet using IP addressing. In order to connect the project you created to a PLC, you need to use the RSLinx software and make sure that the driver you configure in RSLinx matches the IP address of the PLC you are using. You can also connect your PLC to another device such as an HMI.

10. Practical Examples

In this chapter we will bring everything together and go through a couple of example problems. The first problem is a simple tank filling and draining scenario. The second is a relatively simple bottle filling line. Solutions to the problems will be covered step by step. By the end of this chapter you should have a good understanding of how to take a project from scratch through to completion, even if you haven't seen the problem before.

10.1 Tank Filling Scenario

For this problem we will have to keep a tank full enough so that when the system downstream requests water, it can be supplied without overfilling the tank. This system has very few moving parts. It has one inlet valve, one outlet valve and one pump. Water will constantly flow into the tank as long as the inlet valve is open. Water will only flow out of the tank when water has been requested downstream, the outlet valve is open, and the pump is on. Water will flow into the tank quicker than the pump can empty the tank. The tank fill system is a passive part of the overall process, and therefore won't have its own individual start or stop process buttons. The system is shown in the image below.

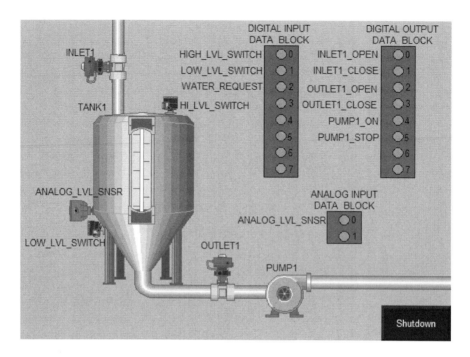

Figure 10-1: The tank filling system

In addition to the tank, valves and pump, the system also includes a low-level switch, high-level switch, and an analog level sensor. The level switches are there as a check for the level sensor in case of failure. The analog level sensor will read the current level of the water in the tank. There is also a display on the tank that shows how much water is in the tank according to the analog sensor (currently the tank is empty). In the image above, all of the inputs and outputs that go to and from the PLC are shown as well.

We now have enough information to create a program to control this system. To start this process, refer back to Chapter 2 and recall the steps needed to properly prepare for creating a program.

It is important to go through the process of creating a flow diagram and planning out how we would like our program to look, before we dive into programming in RS Logix 500. For this example, our flow diagram is provided below.

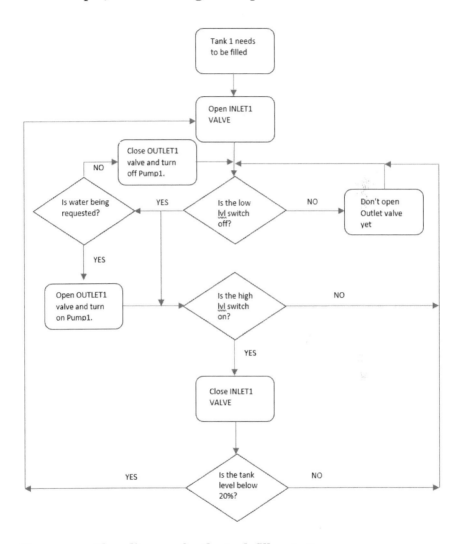

Figure 10-2: Flow diagram for the tank fill system

Using the flow diagram and the IO designations given earlier, we have enough data to start creating our program. For our program, we will set up the program in sections. There will be a fill section of code and a drain section of code.

For the fill section, the program will be set up in the following way. If the tank is below 20% for 0.5 seconds, or the low-level switch turns on, then open the Inlet1 valve to begin filling the tank. If the tank level is above 80% for 0.5 seconds, or the high-level switch turns on, then close the Inlet1 valve. The reason for the timers is to prevent the valve from opening and closing rapidly when the water level is bouncing up and down around the 20% and 80% marks.

In our example the program will be broken up into the INLET VALVE for LAD 2 and the OUTLET VALVE for LAD 3. To control the inlet valve, two timers and two internal data locations will be used. The image below shows the program to control the inlet valve on our tank. The outlet valve will be open if either the low-level switch is on, the water level is low, or if the inlet valve is already open and the high-water level hasn't been achieved.

The inlet valve will close if the water level rises above 80% of the tank level. In case the analog sensor fails, the backup high-level sensor will turn on and the inlet valve will close as well. T4:0 and T4:1 are used as debounce timers set to 0.5s. B3:0.0 is set to turn true when the water level is below 20% for more than 0.5s and B3:0.1 is set to turn true when the water level is above 80% for 0.5s.

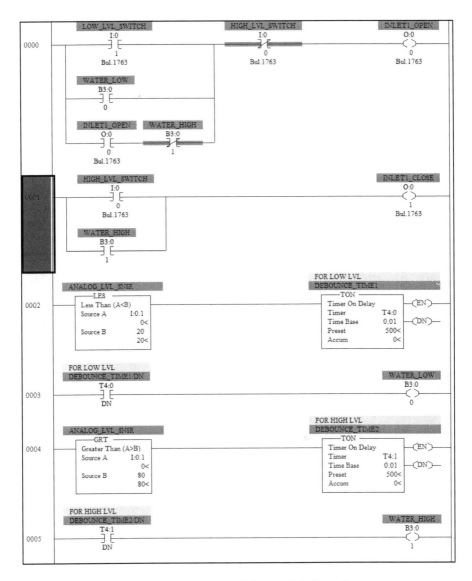

Figure 10-3: The code used to control the tank inlet valve

For the outlet valve, we only want the valve to open when the water level is above 10% of the tank level. The pump should also only turn on when the outlet valve is open. Since this portion of the code controls a pump, we don't want the pump turning on and off intermittently. The pump should run continuously while water is needed, and ideally only turn off when the downstream system no longer wants water.

The pump will turn off when the water level falls below 10%, but won't turn on again until the water is above 20% full. This is to avoid sucking air into the pump. This section of code has two timers and two internal data bits. T4:.2 and T4.3 are used as debounce timers. B3:0.2 tells the outlet valve to close and the pump to turn off. B3:03 notifies that the tank is full enough to begin pumping water. The image below shows the remainder of the tank fill program.

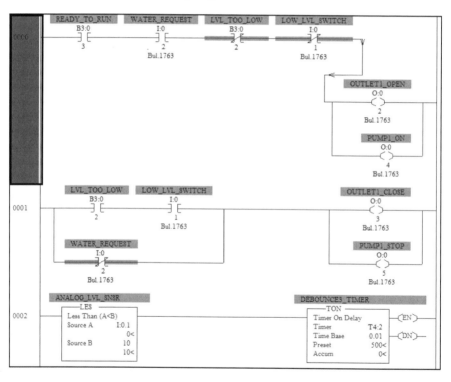

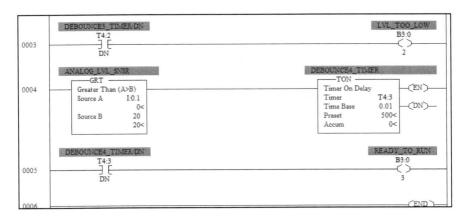

Figure 10- 4 The outlet valve portion of the program.

This example was very simple with only three sensors and three pieces of equipment to control. Hopefully it shows the importance of planning out the program in advance, so that when you build the program you will be able to construct it quickly and effortlessly. The next example will be slightly more complex, but will still be a simple problem.

10.2 Bottling Line Scenario

This example will be of an assembly line where we have a conveyor full of empty bottles that need to be filled. When a bottle is in the correct position, we will begin the fill cycle. Once a bottle is full, it will then continue down the conveyor. There will be a motor to control the conveyor, a sensor to detect when a bottle is present, an extend and retract mechanism for the fill machine, and a sensor to detect when a bottle is full. The image below shows the setup.

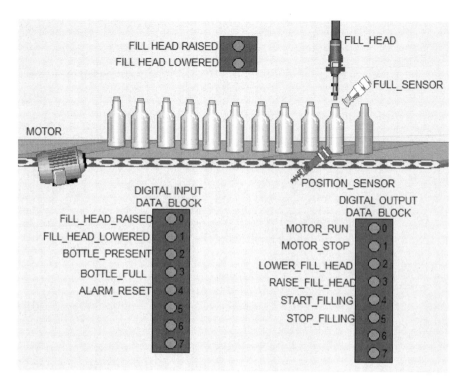

Figure 10-5: A simple bottle filling conveyor system

This is again a relatively simple example which does not have many moving parts. This problem is more difficult however, because more conditions have to be met for a successful fill. For this example, we will not go over the flow diagram. Instead we will discuss the order of operations in enough detail that you should be able to construct your own flow diagram. The process of the conveyor system should go as follows:

1. Make sure the fill head is in the raised position, the conveyor is off, and to check for a bottle at the fill position.

2. If no bottle is at the fill position, turn on the conveyor until a bottle appears (another system will supply the bottles).

3. If a bottle is present, make sure it is full before raising the fill head and starting the conveyor.

4. If the bottle is empty, lower the fill head and start filling the bottle.

5. The conveyor can never run if the fill head is down, and preferably an empty bottle never gets by.

6. It is assumed that all bottles are the same size, don't fall over on the conveyor, and none are broken.

7. It is also assumed that all bottles have been placed far enough apart so that the position sensor turns off before a new bottle moves into the fill position.

8. If the bottle takes more than 10 seconds to fill, then flag an alarm to indicate that the fill head is empty.

In a real-life situation there could be any number of things that could go wrong. The motor could fail, the fill head might not raise up or lower down when told to, a sensor could fail and cause product to be lost. These situations won't be covered in our program, but some will be discussed briefly here.

One alarm could be set up for the fill sensor. It is important when setting up our system, to check how long the average bottle takes to fill up. If we tell the fill head to dispense liquid, it might take two seconds to fill up a bottle under normal circumstances. If a bottle isn't full in 2.5 seconds, we might want to throw an alarm for possible over fill. Another alarm could be set up if we told the conveyor to stop, but the bottle detection sensor keeps going on and off. This would mean that the conveyor is still moving bottles, even though we told the motor to stop. We could also create the opposite if the position sensor is on and we tell the conveyor to move.

For the conveyor portion of the program, the conveyor only runs when the fill position is clear or a bottle is full. The conveyor stops when a new bottle appears at the position sensor. To do this in the code, internal memory bit B3:0.0 is used as the new bottle bit, and bit B3:0.1 is the bottle full bit. When a new bottle is detected, the conveyor instantly stops. If an alarm is active, then the conveyor cannot more. The image below shows the conveyor program in LAD 2.

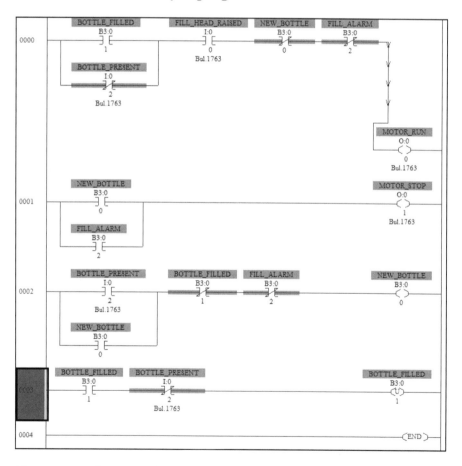

Figure 10-6: The control program for the conveyor motor

138

For the fill head portion of the program, the fill head should default to the raised position. The only time the fill head lowers is when a new bottle is detected and the conveyor has stopped. The fill process begins after the fill head has lowered. If the fill output is true for more than 10 seconds and the full sensor has not been tripped, then an alarm is triggered at bit B3:0.2. Below shows the fill head portion of the program.

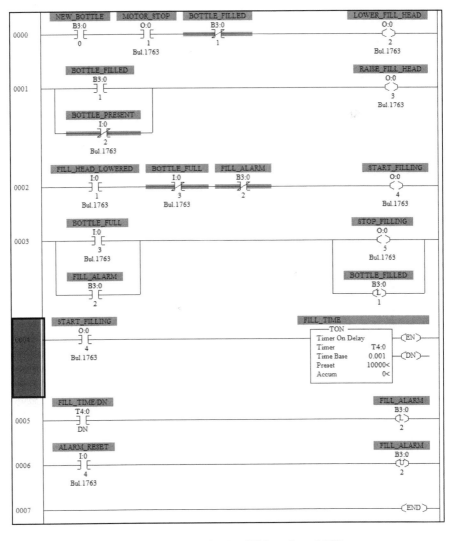

Figure 10-7: Program that controls the fill head and filling sequence

The important takeaways from this project are that even though it is simple, not accounting for possible failures can cause major problems. If the fill head is filling a bottle and no timeout alarm is present, then we could end up filling one bottle until our machine runs out of product. Not to mention that the whole conveyor would be backed up with empty bottles. If you would like to make this problem more challenging, take the above system and try to account for all of the possible alarms.

One important aspect to look for in your program is to make sure that there are no infinite loops, and that every process that moves something can be stopped easily. If you start off by accounting for everything that could go wrong or cause damage to product or equipment, then you will be less likely to overlook potential hazards when equipment starts moving. Before you start programming, make a flow diagram that shows the whole process your program needs to accomplish. Once you are satisfied, start adding symbol names to the bits before you start programming. Also go into each data file and assign a name to everything you plan on using. These methods will help you save time and catch potential errors or missing inputs and outputs.

Further Reading

For further reading on PLC programming, be sure to check out our guide that covers everything you need to know to get started with RSLogix 5000. We help you gain a deeper understanding of the RSLogix 5000 interface, the practical methods used to build a PLC program, and how to download your program onto a CompactLogix or ControlLogix PLC.

**PLC Programming
Using RSLogix 5000**

Understanding Ladder Logic
and the Studio 5000 Platform

a, Available on Kindle and Paperback

Here's a sneak preview of what you can expect:

Introduction to RSLogix 5000 / Studio 5000

We go into meticulous detail on the workings of the Rockwell software, what each window looks like, the elements of each drop-down menu, and how to navigate through the program.

Working with Instructions

We cover every available instruction necessary for beginners, what each instruction does along with a short example for each. You will also learn about communication settings and how to add additional devices to your control system.

Working with Tags, Routines and Faults

We show you how to create and use the various types of tags available, along with all of the different data types that are associated with tags. This guide also covers the finer details of routines, UDTs and AOIs. As well as providing guidance on how to account for typical problems and recover from faults. All of which are essential to most programs.

A Real-World Practical Approach

Throughout the entire guide, we reference practical scenarios where the various aspects we discuss are applied in the real world. We made sure to include numerous examples, as well as two full practical examples, which brings together everything you will have learned in the preceding chapters.

About the Author

Nathan Clark is an expert programmer with nearly 20 years of experience in the software industry.

With a master's degree from MIT, he has worked for some of the leading software companies in the United States and built up extensive knowledge of software design and development.

Nathan and his wife, Sarah, started their own development firm in 2009 to be able to take on more challenging and creative projects. Today they assist high-caliber clients from all over the world.

Nathan enjoys sharing his programming knowledge through his book series, developing innovative software solutions for their clients and watching classic sci-fi movies in his free time.